The NASA STI Program Office . . . in Profile

Since its founding, NASA has been dedicated to the advancement of aeronautics and space science. The NASA Scientific and Technical Information (STI) Program Office plays a key part in helping NASA maintain this important role.

The NASA STI Program Office is operated by Langley Research Center, the lead center for NASA's scientific and technical information. The NASA STI Program Office provides access to the NASA STI Database, the largest collection of aeronautical and space science STI in the world. The Program Office is also NASA's institutional mechanism for disseminating the results of its research and development activities. These results are published by NASA in the NASA STI Report Series, which includes the following report types:

- TECHNICAL PUBLICATION. Reports of completed research or a major significant phase of research that present the results of NASA programs and include extensive data or theoretical analysis. Includes compilations of significant scientific and technical data and information deemed to be of continuing reference value. NASA's counterpart of peer-reviewed formal professional papers but has less stringent limitations on manuscript length and extent of graphic presentations.

- TECHNICAL MEMORANDUM. Scientific and technical findings that are preliminary or of specialized interest, e.g., quick release reports, working papers, and bibliographies that contain minimal annotation. Does not contain extensive analysis.

- CONTRACTOR REPORT. Scientific and technical findings by NASA-sponsored contractors and grantees.

- CONFERENCE PUBLICATION. Collected papers from scientific and technical conferences, symposia, seminars, or other meetings sponsored or cosponsored by NASA.

- SPECIAL PUBLICATION. Scientific, technical, or historical information from NASA programs, projects, and mission, often concerned with subjects having substantial public interest.

- TECHNICAL TRANSLATION. English-language translations of foreign scientific and technical material pertinent to NASA's mission.

Specialized services that complement the STI Program Office's diverse offerings include creating custom thesauri, building customized databases, organizing and publishing research results . . . even providing videos.

For more information about the NASA STI Program Office, see the following:

- Access the NASA STI Program Home Page at *http://www.sti.nasa.gov*

- E-mail your question via the Internet to help@sti.nasa.gov

- Fax your question to the NASA STI Help Desk at (301) 621-0134

- Telephone the NASA STI Help Desk at (301) 621-0390

- Write to:
  NASA STI Help Desk
  NASA Center for AeroSpace Information
  7121 Standard Drive
  Hanover, MD 21076-1320

NASA/TM—2005-212565

EMERGING COMMUNICATION TECHNOLOGIES (ECT) PHASE 4 REPORT

Gary L. Bastin, Ph.D.
ASRC Aerospace Corporation, John F. Kennedy Space Center, Florida

William G. Harris, PE
ASRC Aerospace Corporation, John F. Kennedy Space Center, Florida

José A. Marín
ASRC Aerospace Corporation, John F. Kennedy Space Center, Florida

Richard A. Nelson
NASA, YA-D7, John F. Kennedy Space Center, Florida

**National Aeronautics and
Space Administration**

John F. Kennedy Space Center, Florida 32899-0001

September 2005

Acknowledgments

.Although there is always the risk of inadvertently forgetting someone, the ECT team nonetheless wishes to acknowledge especially the assistance and guidance provided by the following individuals, listed alphabetically. Without the continued support of these supporters who believed in the value of this project, this project could not have accomplished all its goals.

Name	**Organization**
Dr. Larry Andrews	University of Central Florida
Michael Borbath	Harris Corporation
Dr. Geoffrey Burdge	Harris Corporation
Temel Erdogan	ASRC Aerospace Corporation
Bobby Ferrell	NASA KSC
James Kennedy	NASA KSC
Jennifer Murray	NASA KSC
Dr. Ron Phillips	University of Central Florida
Ennis Shelton	NASA KSC
Art Shutt	ASRC Aerospace Corporation
Ed Taff	NASA KSC
Frida Störmqvist Vetelino	University of Central Florida Graduate Student

The help provided by Mr. Ed Taff, NASA Shuttle Landing Facility (SLF) Manager, in coordinating the use of the SLF runway as a laser test range, while not impacting Shuttle Training Activities (STA) required by astronaut crews training for Return to Flight (RTF) is gratefully acknowledged. Mr. James Kennedy, Center Director of the John F. Kennedy Space Center, is gratefully acknowledged in approving our non-standard use of the SLF despite the schedule pressures of RTF. The addressing of safety issues associated with our unusual use of the SLF could not have occurred without the able assistance of Mr. Ennis Shelton, of NASA KSC Health Physics, and of Mr. Arthur Shutt, Safety Engineer of the University-affiliated Spaceport Technology Development Contract (USTDC.) Their continued assistance through the lengthy period of collecting data is gratefully acknowledged. The support of Harris Corporation, in supplying personnel and test equipment to enable the collection of real-time scintillometer data and weather data, is gratefully acknowledged. Likewise, the support of the Florida Space Institute, and of the University of Central Florida, in sharing research equipment and providing assistance from Graduate Students is gratefully acknowledged. The technical support of fSONA Communications Corporation, Richmond, BC, Canada, in answering technical questions and supporting conference calls, is gratefully acknowledged

Available from:

NASA Center for AeroSpace Information
7121 Standard Drive
Hanover, MD 21076-1320

National Technical Information Service
5285 Port Royal Road
Springfield, VA 22161

TABLE OF CONTENTS

No.	Description	Page
1.0	**Introduction**	1
1.1	Background	1
1.2	Objectives	2
1.3	Scope	2
1.4	ARTWG / ASTWG	2
2.0	**Free Space Optics**	5
2.1	Background	5
2.2	Basic FSO Theory	6
2.3	Test Description	7
2.4	Test Objectives	8
2.5	Test Locations & Setups	8
2.5.1	SpaceHab Optical Laboratory	11
2.5.2	EDL East Parking Lot	14
2.5.3	Schwartz Road Test Setup	16
2.5.4	Shuttle Landing Facility	18
2.6	Test Equipment And Software	22
2.6.1	SONAbeam 622-M	22
2.6.2	AirFiber 5800	28
2.6.3	SmartBits	30
2.6.4	Pattern Generators	30
2.6.5	Error Detectors	31
2.6.6	Receiver	31
2.6.7	Clock and Data Recovery (CDR)	31
2.6.8	Weather Station	31
2.6.9	Scintillation	31
2.6.10	SONAbeam Terminal Controller Software	33
2.6.11	SmartApplications	33
2.7	Test Results	33
2.7.1	Prior Art and Historical Rationale for Testing Methodology	33
2.7.2	Testing Methodology	35
2.7.3	Instrument Operation	36
2.7.4	Theoretical Equations	38
2.7.5	Detailed Results	41
2.7.6	Conclusions and Recommendations For Further Research	46
3.0	**EXTENDED RANGE WIFI**	49
3.1	Background	49
3.2	Basic E-R Wi-Fi Theory	49
3.3	Choice Of E-R Wi-Fi For Testing	50
3.4	Test Description	51
3.5	Test Objectives	51
3.6	Test Setup	51
3.6.1	Schwartz Road Test Setup	52

	3.6.2 Shuttle Landing Facility (SLF) Test Setup	54
3.7	Test Equipment And Software	54
	3.7.1 DragonWave AirPair 100	54
	3.7.2 Laptop	59
	3.7.3 DragonWave AirPair 100 Controller Software	59
	3.7.4 DragonWave Link Tool	62
	3.7.5 Throughput Testing	63
3.8	Test Results	64
	3.8.1 General	64
	3.8.2 Results Summary	65
3.9	E-R Wi-Fi Security Concerns	68
3.10	E-R Wi-Fi Summary And Recommendations	68
4.0	**Ultra Wide Band**	**71**
4.1	Introduction	71
4.2	Background	71
4.3	UWB Waveform INVESTIGATION UPDATE	72
4.4	Standards updates	73
4.5	UWB antenna updates	73
4.6	Summary of Results	74
4.7	Recommendations for Future Research	74
5.0	**ECT Summary Recommendations For Continued Research**	**75**
APPENDIX A:	**Acronyms & Definitions**	**77**
APPENDIX B:	**Optical BER Equations For OOK**	**81**
APPENDIX C:	**Extended Range Wi-Fi Test Results**	**83**

LIST OF FIGURES

No.	Description	Page
Figure 1-1	ARTWG National Development Strategy	4
Figure 2-1	Typical Link-Verification Setup Using SmartBits	10
Figure 2-2	Typical BER Test Setup	10
Figure 2-3	SpaceHab Optical Lab Setup	12
Figure 2-4	fSONA Units Being Tested In The SpaceHab Optical Lab	12
Figure 2-5	Beam Measurements Between OTU #1 & OTU #2	13
Figure 2-6	EDL Parking Lot	14
Figure 2-7	OTU-1 in EDL Parking Lot	15
Figure 2-8	OTU-2 At South End of EDL Parking Lot	15
Figure 2-9	AirFiber #1 On Trailer #1 in EDL Parking Lot	16
Figure 2-10	OTU #2 Parked In The Bend On The West End of Schwartz Rd.	17
Figure 2-11	OTU #1 Positioned @ 1 Km On Schwartz Rd.	17
Figure 2-12	SLF Viewed From North	18
Figure 2-13	Typical Aircraft Traffic at SLF	19
Figure 2-14	fSONA #2 on SLF Centerline	19
Figure 2-15	fSONA #1 With SmartBits, Laptop & Open Control Panel	20
Figure 2-16	SLF South Setup With fSONA #1, FSI Trailer, and Scintillometer	20
Figure 2-17	South SLF Setup With Van and Trailer	22
Figure 2-18	Front of SONAbeam 622-M OTU	23
Figure 2-19	SONAbeam Junction Box Open	25
Figure 2-20	Pole Mount with OTU, Connection Box & Power Supply Box	25
Figure 2-21	Aperture Adapter and Mask	26
Figure 2-22	Aperture Adapter and Mask	27
Figure 2-23	AirFiber 5800 With Fiberglass Shroud Removed	28
Figure 2-24	Field Testing with the SmartBits & a Laptop	30
Figure 2-25	Wx Station and Scintillometer Rx on SLF	32
Figure 2-26	Transmit Portion of Scintec Scintillometer	32
Figure 2-27	Measured Average C_n^2 Values	36
Figure 2-28	Values of Inner Scale from 1-min. Avg. Wind Speed	37
Figure 2-29	1-min. Avg. BER Vs. SNR	42
Figure 2-30	5-min. Avg. BER Vs. SNR	43
Figure 2-31	BER & Error Count Vs. Time of Day & Aperature Dia.	44
Figure 2-32	BER Vs. Time of Day @ 1.0 & 1.25 Aperture Dia.	45
Figure 2-33	BER & C_n^2 Values Vs. Time of Day @ 1.0 & 1.25 Aperture Dia.	46
Figure 3-1	AirPair 100 On Trailer #2 At West End Of Schwartz Rd.	53
Figure 3-2	AirPair 100 On Trailer #1 At East End Of Schwartz Rd.	53
Figure 3-3	Modified AirPair 100 Radio with 30cm Antenna & Added Riflescope	55
Figure 3-4	AirPair 100 Outdoor Modem	56
Figure 3-5	Power + Ethernet Cable	58
Figure 3-6	Pole Mount with AirPair 100 Modem, Radio, and Antenna	58
Figure 3-7	Laptop connected to the AirPair 100 Modem	60

Figure 3-8   CLI screen capture (Laptop) 61
Figure 3-9   Web Interface – General Radio Information Page 61
Figure 3-10   LinkTool utility 63
Figure 3-11   Throughput Versus Distance 67
Figure 3-12   Average Throughput versus Distance 68

LIST OF TABLES

No.	Description	Page
Table 2-1	FSO Industry Comparisons for FY05	6
Table 2-2	FSO Test Locations	11
Table 2-3	SONAbeam 622-M Specifications	23
Table 2-4	AirFiber 5800 Specifications	29
Table 2-5	Summary of Test Locations & Objectives	33
Table 3-1	Test Locations	52
Table 3-2	AirPair 100 Specifications	57
Table 3-3	AirPair 100 Test Configuration Parameters	59
Table 3-4	Optimum RSL Values Vs Distances	62
Table 3-5	Optimum Parameters	62
Table 3-6	Summary of Test Locations & Objectives	65

Executive Summary

The National Aeronautics and Space Administration (NASA) is investigating alternative approaches, technologies, and communication network architectures to facilitate building future Spaceports, building on the communications capabilities that already exist. These investigations support the development of communication networks for use with the Crew Exploration Vehicle (CEV), as well as other crafts presently under development or under consideration in the Government, academic, and private sectors. Forming part of the associated Ground Support Equipment (GSE) that will be necessary to prepare and launch these space crafts, the communication technology resulting from these investigations will also provide a national centralized R&D forum for next-generation Spaceport and Range technology development. Together, these sectors share the common goal of changing the historic risk/reward equation for access to space, especially with regards to communications networks, with the intent to:

- Dramatically reduce launch cost
- Greatly improve launch system reliability
- Significantly reduce crew risk

During FY05, ECT Phase 4 accomplished the following major milestones:

- Performed FSO scintillation and aperture averaging testing on the Shuttle Landing Facility (SLF) Runway, with the runway used as a giant laser test range
- Validated published theoretical equations for predicting the performance enhancements possible through increasing receiver apertures to mitigate the effects of optical scintillation
- Wrote and published an SPIE Technical Conference Paper, San Diego (August 2005) on the use of aperture averaging to mitigate scintillation and improve Bit Error Rate (BER) for high data rate Free Space Optical links
- Supported and encouraged University of Central Florida personnel to utilize data through sharing SLF optical test data, which resulted in two additional technical conference papers being published at the SPIE in San Diego (August 2005)
- Trained personnel on extended range WiFi equipment at vendor facility, and subsequently performed extended range WiFi testing within the unique Kennedy Space Center environment
- Evaluated and tracked UWB industry developments
- Identified 3 additional technology topics needing investigation in Phase 5, FY06: (WiMAX, second generation UWB, and FSO networking and beam wander issues)

1.0 INTRODUCTION

1.1 BACKGROUND

Emerging Communication Technologies is a multi-year task investigating new communication technologies with a high probability of utilization and application for future ranges and spaceports.

In year one, the project was called Range Information Systems Management (RISM)[1]. This project investigated US ranges and documented their missions, capabilities, and infrastructures. A part of this investigation was the identification of past historical trends in communication technology and the identification of new, emerging technologies which might offer improved range communication capabilities in the near future. Three emerging communication technologies were identified: Free Space Optics (FSO), Extended Range Wireless Ethernet (Wi-Fi), and Ultra Wide Band (UWB). All three of these technologies address the first mile / last mile communication solution.

In year two, specific examples of FSO, UWB and Wi-Fi were purchased and evaluated for range application. The FSO hardware purchased and evaluated was an AirFiber 5800 optical transceiver. This unit includes an auto-tracking feature that maintains optical alignment during small movements that normally occur to support structures due to diurnal solar heating, winds, and vibration. The units were tested over various distances and through various weather conditions.

The UWB effort in year 2 produced an industry survey, developed a fundamental basic UWB mathematical theory, and included testing an Evaluation Kit (EVK) from Time-Domain Corporation. Signal degradation due to range; normal office barriers of concrete, metal, partitions, etc.; and the effects of interference from microwave ovens, wireless phone, etc., were measured.

During year 2, two different Wi-Fi systems were purchased and evaluated. One was an 802.11b base station from Microsoft. The other was an 802.11g system from D-Link. The majority of testing for the 802.11b system were signal degradation due to range; normal office barriers of concrete, metal, partitions, etc.; and interference from microwave ovens, wireless phone, etc.

During year 3, wide-beam, non-tracking FSO hardware from fSONA was procured and evaluated in contrast to the narrow-beam, auto-tracking hardware from AirFiber investigated in year 2. Additionally, OFDM-UWB performance limitations were investigated through using the same EVK with the addition of new firmware upgrades from Time-Domain Corporation as a test bed. Likewise, general industry trends for achieving first mile / last mile communications were monitored and incorporated into this multi-year ECT activity.

---

[1] Acronyms and Definitions are provided in Appendix A.

During year 4, the current year, extensive investigations into FSO, Wi-Fi, and UWB technologies continued. Extended-Range (E-R) WiFi technology investigation and evaluation was added to the project. Original plans were to investigation performance impacts due to multiple FSO wide-beam systems operating in close proximity; however, budget constraints and the delayed receipt of funds prevented purchasing a second FSO system. With only one wide beam FSO system available, planned FSO scintillation testing was accelerated through joining the University of Central Florida, Florida Space Institute, Technical University of Catalonia (Spain), and Harris Corporation in investigating aperture averaging and scintillation effects on Bit-Error-Rate. Jointly conducted testing and theoretical development resulted in three technical papers that were presented in August at the San Diego SPIE 2005 Conference. The data for these papers were based on multiple FSO tests conducted on the Shuttle Landing Facility (SLF) runway. E-R WiFi investigations centered around a Dragonwave COTS system procured and then tested over various distances around Kennedy Space Center. UWB activities during year four were limited to monitoring developments within the industry, commencing a study of UWB waveforms optimized for efficiency, and waiting for new evaluation kits to be released. Likewise, general industry trends for achieving first mile / last mile communications were monitored and incorporated into this multi-year ECT activity.

1.2 OBJECTIVES

The primary objective for the Emerging Communication Technology (ECT) task is to lead the development of a Space Based Range Distributed Subsystem (SBRDS) network that provides the concurrent features and growth capabilities necessary for future Spaceports and Ranges to interconnect Range assets, Range operations, and Range users during launch and recovery events, while focusing primarily on the First Mile/Last Mile wireless communication extensions to existing, fixed communication infrastructures.

1.3 SCOPE

ECT Phase 4 was limited to the following:
- Help develop future range and spaceport architectures and needs
- Investigate FSO scintillation and aperture averaging effects using a fixed alignment, wide-beam FSO system
- Procure, test, and evaluate an Extended Range WiFi system
- Investigate the capabilities of the Time-Domain upgraded firmware UWB system
- Commence investigating optimal UWB templates for improving overall system power efficiency.

1.4 ARTWG / ASTWG

ECT helped develop future range and spaceport architectures and needs through participation in the active efforts of:

- ARTWG (Advanced Range Technology Working Group)
- ASTWG (Advanced Spaceport Technology Working Group)
- FIRST (Future Interagency Range & Spaceport Technologies)

ARTWG is a collaborative NASA/US Air Force/Industry/Academia effort to focus interest and investment in Range technologies. It is co-chaired by NASA and the US Air Force, and consists of aerospace leaders from industry, academia, and national, state, and local governments. ARTWG addresses Range development needs while its companion organization ASTWG (Advanced Spaceport Technology Working Group) addresses Spaceport development needs. . The initial ARTWG roadmap for future ranges was released in March 2004 (Figure 1-1).

Figure 1-1   ARTWG National Development Strategy

FIRST is the Future Interagency Range and Spaceport Technologies consortium. FIRST is a Tri-Agency consortium consisting of NASA, FAA, and DoD. Its goal is to enable joint planning of spaceports and range technologies with each agency controlling its own development monies.

ARTWG, ASTWG and FIRST complement ECT activities and provide opportunities to interact with Government, industry, and academic personnel on new developments applicable to Ranges and Spaceport of the future. National conferences are held once or twice a year. During FY-05, ECT personnel attended the joint ARTWG / ASTWG conference in Colorado Springs, CO between January 10th and January 14, 2005.

2.0   **FREE SPACE OPTICS**

2.1   **BACKGROUND**

Free Space Optics (FSO) was one of the original three First Mile/Last Mile broadband wireless access systems identified in the RISM Phase 1 report[2]. An auto-track system from AirFiber was purchased, tested and reported during the ECT Phase 2 activities[3]. For ECT Phase 3, a fixed-alignment, wide-beam, multiple transmit beam system by fSONA was procured and tested. In Phase 4, this system was further utilized to investigate scintillation and aperture averaging effects

FSO is a maturing technology that offers significant advantages over most wireless technologies, including higher data rate, and the complete avoidance of any spectrum licensing costs. Optical communication systems provide the highest available carrier frequencies and thus the fastest data rates possible today. FSO is designed to be a lower cost alternative to conventional fiber-optic cable-based communication links[4]. FSO is especially attractive within metropolitan environments where the costs for trenching, cable installation, and street repairs can run from $200K to easily over $1M per mile, depending on the urban location.

Although FSO offers the potential of maximum wireless performance, the FSO industry continues to be reshuffled. Table 2-1 lists and compares key international FSO players. AirFiber, the industry leaders 3 years ago is presently out of business. Terabeam, another top FSO company just a few years ago, has sold their assets to Harris Corporation and has gone out of the FSO business, focusing instead on wireless opportunities.

The primary limitations for FSO involve weather and atmospheric effects. Since ground links traverse horizontally through the lowest portion of the atmosphere, weather and the atmosphere tend to magnify attenuation and scintillation effects. Fog is the primary weather concern especially at a 1 mile or less distance. Scintillation effects usually do not create serious problems for "First Mile" low to medium data rate links; however, they have significant impact on the reliability of higher data rate links and longer links, especially.

---

[2] Range Information Systems Management (RISM) Phase 1 Report, NASA/TM-2004-211523, September 2002
[3] Emerging Communication Technologies (ECT) Phase 2 Report, Volume 1, Main Report, NASA/TM-2004-211522, September 2003
[4] http://www.airfiber.com/products/index.htm

Table 2-1   FSO Industry Comparisons for FY05

Company	Tx	Auto Track	Wave Length	Comments
AirFiber	1	Y	785	Out of business
Alcatel SA	-	-	-	Uses fSONA equipment
Cablefree Solutions	3	N	780	UK
Canon Inc.	1	Y	785	USA
Communication By Light GmbH (CBL)	4	N	870	Germany
Corning Cable Sys	4	Y	850	USA
Dominion Lasercom	1	Y	850	USA
fSona Com	4	N	1550	Canada
iRLan Ltd.	1	N	unk	Israel
LaserBit Com	8	N	unk	USA.
LightPointe Comm	4	Y	850	USA.
LSA Photonics	1	N	785	USA
Maxima Corp	unk	unk	10,000	Long wave length infrared; USA
Mostcom Ltd	-	-	-	Same as Sceptre
Omnilux Inc.	3	unk	unk	Mesh design, USA
Sceptre Comm. Ltd.	2	N	850	UK & Russia
Terabeam Corp.	*1*	*Y*	*1550*	*Out of the FSO business, FSO Intellectual Property bought by Harris Corp.*
TereScope	1	Y	850	USA

FSO testing during year 4 of the ECT project was conducted using the fSONA SONABEAM 622-M transceiver units. Some limited confirmational testing was accomplished with the AirFiber units purchased previously, during year two.

2.2   BASIC FSO THEORY

Free Space Optical (FSO) communication was discussed at length in the previous Phase 1 RISM report and the ECT Phase 2 report. A brief summary is repeated here for continuity, and to serve as an introduction to the technology for anyone unfamiliar with the technology.

FSO dates to pre-history. Extensive FSO networks were established in the 19th Century throughout France and North Africa, based on semaphore systems. Later, during the latter part of the 19th century, FSO telephone communication was developed.

The modern FSO age commenced with the invention of the laser slightly more than 45 years ago. Generation of wavelength stable coherent light at selected wavelengths provided the ability to select specific fixed wavelengths to achieve FSO systems that enable lessening atmospheric attenuation, provide operation through rain, and achieve eye-safety.

Fundamentally, modern FSO systems typically employ NRZ (non-Return to Zero) modulation of laser light. Digital data is encoded as either a high intensity beam or as a low intensity beam, separated by the extinction ratio present in the ON to OFF states engendered by the modulating device.

Within the receiver, a photodetector provides the optical to electrical (OE) conversion. Depending on the range over which communication is desired, both Positive-Intrinsic-Negative (PIN) diodes and APDs (Avalanche Photodetector Diodes) are used. PIN diodes provide less sensitivity, but require only minimal voltage bias to make them operational. APDs provide the maximum performance in sensitivity, but require voltages often exceeding 100 Volts dc to achieve this high sensitivity. This, in turn, increases the need for properly coating circuit cards for FSO apparatus intended for use outdoors, to avoid issues with condensing moisture causing short-circuits.

At the output of the photodetector is a Trans-Impedance Amplifier (TIA). Its purpose is to provide the necessary gain by which to generate a voltage from the current produced by the photodetector diode when exposed to light. Beyond this lie the framing and other packetizing electronics, needed to provide the proper data interfaces for the subsequent parts of the communication system. For fiber optic extensions, it is necessary to have clock and data recovery circuitry, by which clocks are derived from incident light pulses coming into the FSO system via fiber optic cable, to provide proper timing interfaces.

For the fSONA system tested on this project, a Smartbits OC-12 fiber optic interface operating at 622 Mb/s served as the physical data interface in and out of the two OTUs. Advantest and Anritsu pattern generators and error detectors were also used at times for performing these same duties during SLF testing.

2.3 TEST DESCRIPTION

Phase 4 testing with FSO units included two FSO systems, tested at four locations over various distances. Most of the FY05 testing was performed with the fSONA 622-M system. A short test was also run using the AirFiber units procured during Phase 2. Test locations included the following:

- SpaceHab optical laboratory
- EDL building rear parking lot
- Schwartz Road
- Shuttle Landing Facility (SLF)

2.4 TEST OBJECTIVES

The FSO test objectives were as follows:

- Evaluate COTS FSO equipment for possible future use at KSC
- Identify any fundamental shortcomings that must be filled in commercial FSO communication technologies prior to integrating these technologies into future range architectures
- Compare a fixed alignment, multiple beam, wide beam system against the previously-tested auto-tracking, single beam, narrow beam FSO system
- Investigate scintillation and other atmospheric limitations on FSO, and assess aperture averaging mitigation on Bit Error Rate (BER)

2.5 TEST LOCATIONS & SETUPS

Test setups varied by location and equipment. A pair of antenna trailers were modified during earlier phases of ECT to enable quick mounting of either the fSONA or AirFiber FSO systems. These trailers were used for all outside testing and enabled quick positioning of the FSO systems. A pair of portable stands were used for indoor testing of the fSONA units.

For outdoor testing, the basic test setup was to position the two trailers at a fixed distance, align the Optical Transfer Units (OTU), establish a data link, test the quality of the link and then conduct any scintillation and aperture averaging mitigation measurements.

Initial alignment of the fSONA system was accomplished using a 12X power Leupold™ rifle scope. This gross alignment was usually adequate to establish an initial, but not optimally aligned link. The link was then electronically fine-tuned using the factory provided software on a laptop computer. This software provides an output display of the receiver's incoming energy in microwatts. Mechanical adjustment was accomplished using jacking screws on both azimuth and elevation lever arms. Once the alignment was completed, the azimuth and elevation axes were locked and not changed further during testing.

The AirFiber units were aligned using their internal cameras and their self alignment feature. The OTU were aligned by mechanically positioning each OTU with its target OTU in the camera's field of view. The internal crosshairs were driven to center on the peer OTU and then the system's auto-alignment feature was activated. The AirFiber system then automatically located, centered, and maintained both OTUs for an optimum link.

A SmartBits Bit Error Rate test device was used to test and verify the initial quality of the fSONA link. Early during testing, it was discovered that establishing a link based only the receive power and the fSONA status indicator was not adequate to ensure a solid link.

Instead, we found it necessary for the SmartBits to measure throughput and packet loss for various packet sizes to ensure a solid link. This setup is shown in Figure 2-1. The SmartBits sent a variable length data stream to the input of OTU #1. OTU #1 transmitted the data packets to OTU #2 using the FSO laser. OTU #2 received the FSO data packets, and converted these to a multi-mode fiber-optic output. This output was looped back to the input of OTU #2 through a short fiber optic cable jumper. The data packets were then converted back to another FSO signal and sent back along the same propagation path from OTU #2 to OTU #1. The fiber output from OTU #1 was directed back to the SmartBits where a comparison was made between the transmitted and received packets to determine any throughput errors or packet losses.

After the initial link was established and verified with the SmartBits, additional equipment was used to measure the Bit Error Rate (BER) with more flexibility in the choice of specific $2^N - 1$ maximal length test sequences than possible with the SmartBits, setting $N = 23$ (i.e., PRBS 23), typically, although setting $N = 21$ and lower was also used on occasion for trouble-shooting test configurations. A typical test setup is shown in Figure 2-2. Specifically, an Advantest D3186 Pattern Generator was used to input a test pattern (usually PRBS 23) to OTU #2. This data pattern was sent via FSO to OTU #1 using only one of the four laser transmitters (Tx). The fiber output from OTU #1 was then routed to an Agilent 83440C Receiver that converted it to an electrical signal. The electrical signal was then sent to a Mindspeed M21012 Clock & Data Recovery (CDR) unit. The data and clock outputs from the CDR went to an Advantest D3286 Error Detector where the BER measurements were made. For some tests, an Anritsu MP1763C Pattern Generator and an Anritsu MP1764A Error Detector were substituted for the Advantest equipment. Measured BERs would vary by test as the receive area of fSONA #1 was reduced using aperture masks and scintillation varied due to diurnal variation of solar heating of the SLF. Masks and test equipment are discussed in more detain in a later section. For Scintillation testing, the setup also included a Scintec BLS900 Scintillometer and Cambell Scientific Weather Station.

ASRC Aerospace personnel ran the fSONA and SmartBits equipment, and performed alignment of the FSO links. FSI personnel provided and ran the pattern generators, error detectors, weather station and scintillometer.

The choice of test location was selected based on the purpose of the test and the distance to be investigated. The SpaceHab Optical Lab was used for very short distance laboratory testing. The EDL parking lot was used to verify test setups prior to deploying into the field for longer distances. Schwartz Road was used for distances up to 2.5 miles when scintillation effects were not being investigated. The SLF was utilized for its controlled environment while doing 1 km testing of scintillation effects. A summary of test locations and distances is included in Table 2-2. Testing was not performed in any specific order. Details about each test location are included in the following sections. The FSO units, test equipment and software are described in later sections.

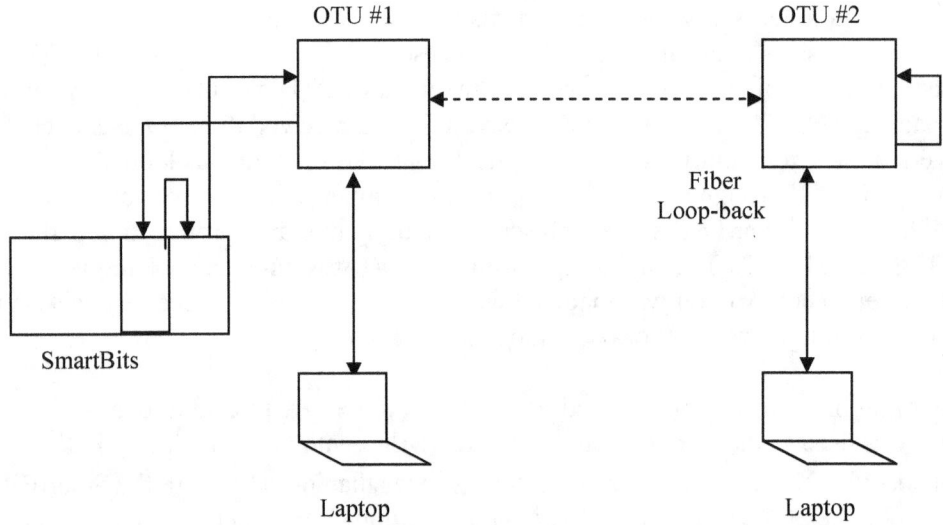

Figure 2-1   Typical Link-Verification Setup Using SmartBits

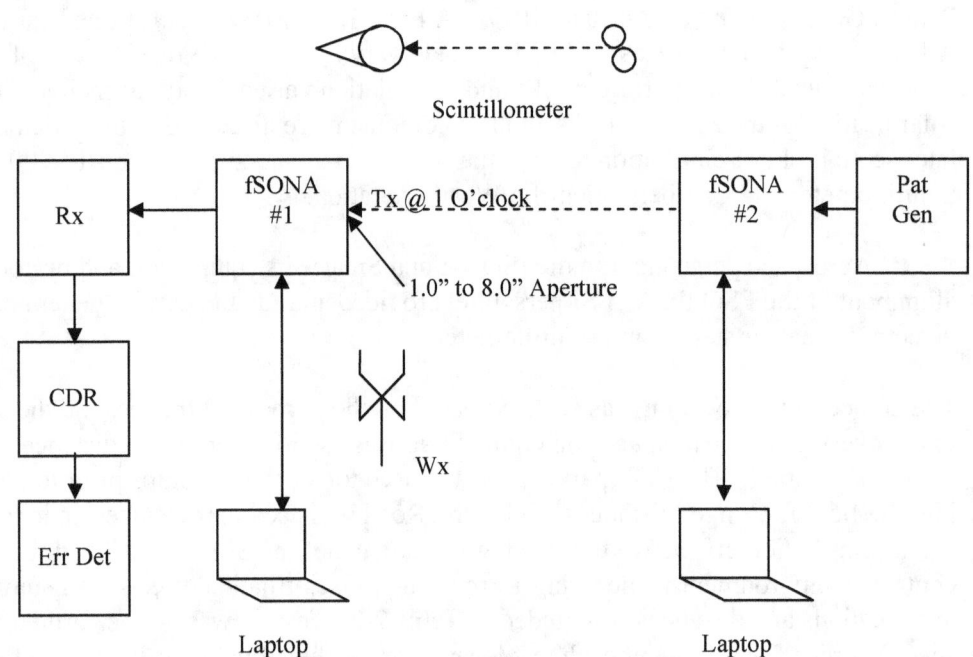

Figure 2-2   Typical BER Test Setup

Table 2-2   FSO Test Locations

Location	One Way Distance	Loop Back
SpaceHab Optics Lab	10 ft	N
EDL East Parking Lot	175 ft	Y
Schwartz Road	1.0 to 2.5 miles	Y
SLF	1000 km	Y/N

2.5.1   SpaceHab Optical Laboratory

SpaceHab is a payload processing facility owned by AstroTech and located just outside the Cape Canaveral Air Force Station south gate. Office space and facilities are also leased to others in the space industry. The Florida Space Institute (FSI), a part of the University of Central Florida (UCF), maintains an optical laboratory at this facility.

The fSONA units were removed from their trailers and taken to the FSI Optical Laboratory at SpaceHab for precise testing and characterization. The fSONA units were mounted on their short support stands that were fabricated during Phase 3. These stands were placed on a 10-foot optical table inside the environmentally controlled optical lab. During the subsequent tests, only a single transmitter on OTU #1 was used as the signal source. Due to the close proximity of the OTUs, calibrated attenuators were mounted to the exterior of the 1 O'clock transmitter. The other three transmitters were turned OFF. Figure 2-3 shows the test setup within the lab. Figures 2-4 and 2-5 provide additional views of the laboratory testing and some of the test equipment. Newport and Ophir power-meters were used during the tests. One series of tests measured receive power at OTU #2. Another series measured the laser beam power and signal at a position between OTU #1 and OTU #2. FSI research associates from Harris Corporation performed these measurements.

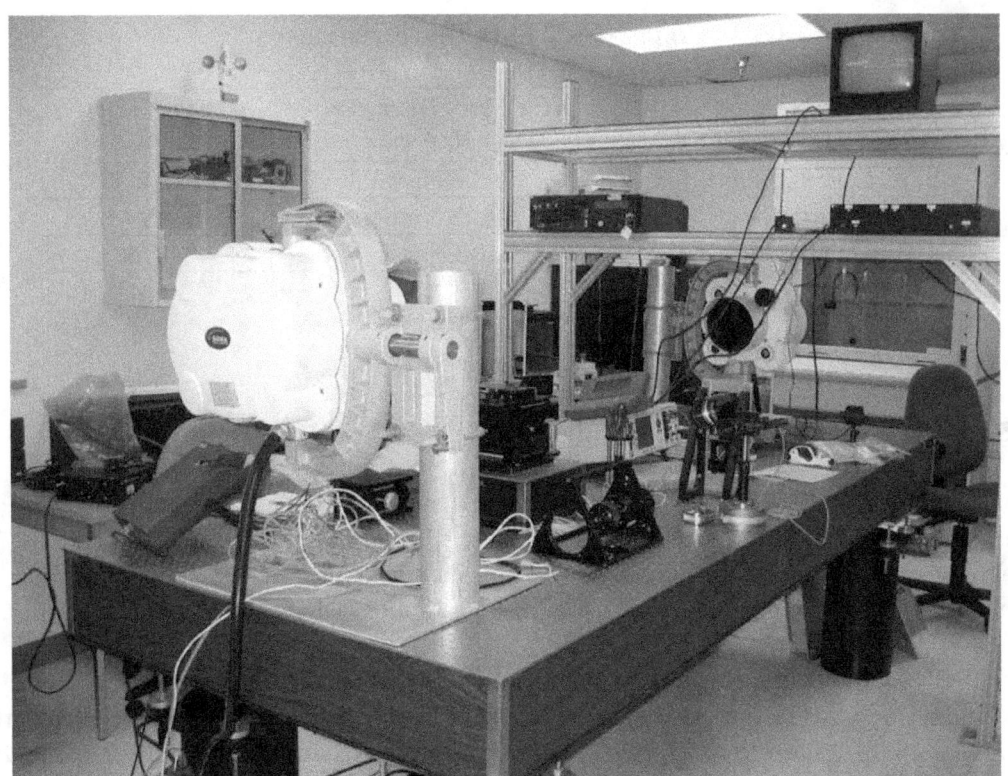

Figure 2-3 SpaceHab Optical Lab Setup

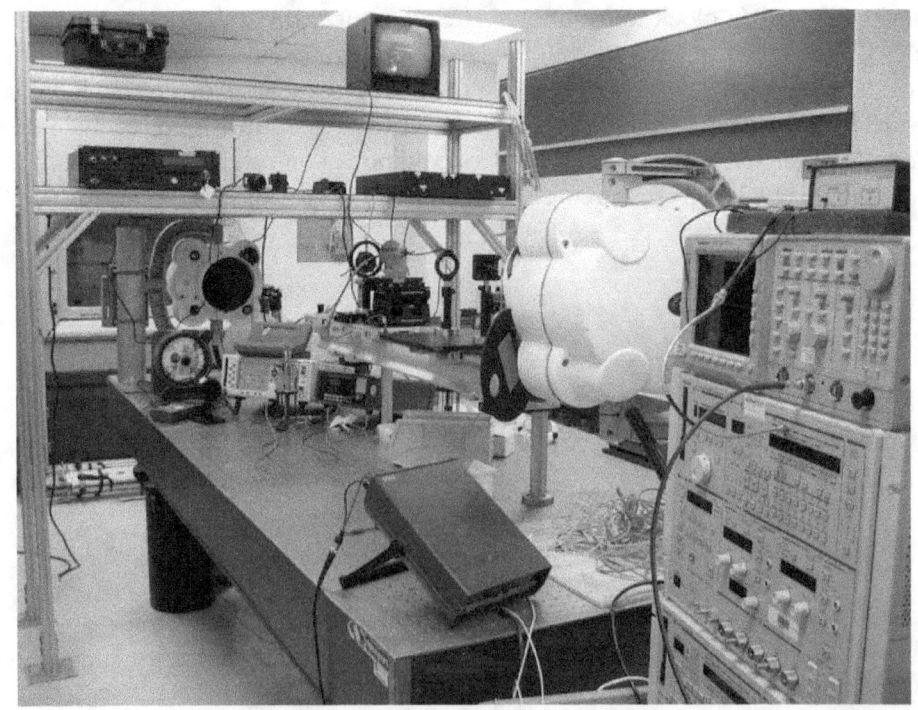

Figure 2-4 fSONA Units Being Tested In The SpaceHab Optical Lab

Figure 2-5   Beam Measurements Between OTU #1 & OTU #2

2.5.2 EDL East Parking Lot

The EDL East parking lot is a convenient location for conducting initial shake-down, procedural, and other preliminary tests where long distances are not required (Figure 2-6). Trailers are usually positioned about 175 feet apart. Figure 2-7 shows fSONA #1 mounted on Trailer #1 located on the North end of the parking lot (left). Figure 2-8 shows Trailer #2 & fSONA #2 on the South end of the parking lot (right). Portable generators are used for power during these tests. The tests were usually set up per Figure 2-1.

Figure 2-6   EDL Parking Lot

Figure 2-9 shows both the AirFiber and fSONA units mounted on Trailer #1. The AirFiber units were tested to verify they were operational and to compare and contrast their auto-alignment process with the fSONA's manual alignment process. The AirFiber and fSONA OTUs could not be simultaneously used on the trailers because the signal to fSONA #1 on Trailer #1 would be blocked by AirFiber #1.

Figure 2-7   OTU-1 in EDL Parking Lot

Figure 2-8   OTU-2 At South End of EDL Parking Lot

Figure 2-9   AirFiber #1 On Trailer #1 in EDL Parking Lot

2.5.3   Schwartz Road Test Setup

Schwartz Road is a remote East-West road at KSC that runs relatively straight for 2.7 miles. During Phase 3, extensive long-distance testing of the fSONA units was performed on Schwartz Road. Some additional testing was also performed at this location during Phase 4. This testing was primarily run to verify the baseline values of the fSONA units prior to SLF testing. Trailer #2 with fSONA #2 was placed in the 90-degree bend on the West end of Schwartz Road (Figure 2-10). Trailer #1 was positioned off to the side of the road at various distances (Figure 2-11). Initial testing was performed at 1.0 mile. This was followed with practice tests at 1.0 km. The baseline test configuration was again per Figure 2-1 with the SmartBits and Laptop positioned on trailer #1 and a fiber loop-back installed at OTU #2. Power was provided by a portable generator. BER tests were per Figure 2-2 without the weather and scintillation equipment. All equipment was run off a single portable ac generator at each end.

Figure 2-10 OTU #2 Parked In The Bend On The West End of Schwartz Rd.

Figure 2-11 OTU #1 Positioned @ 1 Km On Schwartz Rd.

2.5.4 Shuttle Landing Facility

All scintillation testing was performed under the controlled environment of the Shuttle Landing Facility (SLF). Its controlled shape, orientation, and uniformity make it an ideal laser test range. Figure 2-12 is an aerial view from the North end of the SLF. The VAB is visible near the top left. General specifications for the SLF are as follows:

- 15,000 feet active runway
- 300 feet wide
- Precisely grooved and milled surface
- Orientation: 150 / 330 Degrees

The SLF is an active airport when the Space Shuttle is not landing (Figure 2-13). This limits the days and hours of testing. Special rules and safety training are also required.

The testing setup at the SLF followed the same configuration at the other sites. The trailers with the fSONA units onboard were positioned 1000 meters apart on or near the centerline of the SLF. Unit #1 was always on the South side (Figures 2-14 & 2-15) and Unit #2 on the North side. Unit #1 was always located at the 3000-foot mark referenced from the South runway edge (concrete portion) of the SLF. The 3000-foot location was selected since the first 3000 feet have had the rain grooves removed. The fSONA units were usually placed on the centerline of the SLF except for some later tests that included a pair of FSI equipment trailers. For these latter tests, the FSI trailers were placed on the centerline and the fSONA units were moved 20 feet east of the centerline (Figure 2-16).

Figure 2-12   SLF Viewed From North

Figure 2-13   Typical Aircraft Traffic at SLF

Figure 2-14   fSONA #2 on SLF Centerline

Figure 2-15 fSONA #1 With SmartBits, Laptop & Open Control Panel

Figure 2-16 SLF South Setup With fSONA #1, FSI Trailer, and Scintillometer

Once the trailers were in place, the units were aligned both mechanically and electronically. The mechanical alignment was accomplished via a 12X riflescope using jacking screws on both azimuth and elevation axes. Electronic alignment employed using the same jacking screws to obtain a peak receive power using the factory software. During latter tests, alignment was based on peak receiver voltage at the output ports inside the control panel (Figure 2-15). The voltage was read using a digital multi-meter. This method tended to provide a quicker readout with less noise. The verification setup was as previously shown in Figure 2-1. Tests using the SmartBits, seen under the laptop in Figure 2-15, confirmed a good link was established. Throughput values of 100% and packet losses of 0% were usually achieved for all packet sizes.

Alignments were performed early in the day prior to the presence of the worst of the day's scintillation using a loop-back link from the far-end to the near-end, traversing the 1 km span twice. Once aligned, one-way scintillation tests were then conducted through local solar noon, collecting measured data of BER versus receive aperture diameters.

BER testing was then performed in conjunction with the University of Central Florida and the Florida Space Institute per Figure 2-2. The BER measurement equipment was inside the FSI trailer shown to the left in Figure 2-16 or inside the van shown in Figure 2-17. The scintillometer receiver is visible under the left edge of the awning in Figure 2-16.

Figure 2-17   South SLF Setup With Van and Trailer

2.6   TEST EQUIPMENT AND SOFTWARE

Key FSO test hardware included the following:

- SONAbeam 622-M - Optical Transceiver Unit (OTU)
- AirFiber 5800 OC-12 - Optical Transceiver Unit (OTU)
- SmartBits – Data packet source for measuring Throughput and Packet Loss
- Advantest D3186 – 10 Gbps Pattern Generator
- Advantest D3286 – 10 Gbps error Detector
- Anritsu M1763C – 10 Gbps Pattern Generator
- Anritsu M1764A – 10 Gbps Error Detector
- Agilent 83440C 20 GHz Receiver
- Mindspeed M21012 - Clock and Data Recovery
- Campbell - Scientific Weather Station
- Scintec BLS900 - Scintillometer (Transmitter & Receiver)

In addition to the above hardware, software packages were instrumental in testing and data acquisition. These software packages included:

- SONAbeam's STCv3 - Terminal Control Software
- AirFiber's CamLAP – Link Acquisition Program
- SmartBits's SmartApplications – Packet Testing Program
- LabView BER calculations & Data Acquisition
- Scintec operations software
- Campbell Scientific Wx Station operations software

2.6.1   SONAbeam 622-M

The SONAbeam 622-M OTU, shown in Figure 2-18, was the primary component under test. Specifications for the units are summarized in Table 2-3. A pair of 622-M were purchased around 3/1/04 under the ECT task order (#00087), Phase 3.

The SONAbean 622-M is a fixed alignment, multi-beam design. Initial alignment is normally achieved using an eye-safe unfiltered 12x power rifle scope at longer distances, and a lower-power eye-safe filtered 9x rifle scope at shorter distances. Fine alignment is achieved using the fSONA software or a voltmeter.

Figure 2-18   Front of SONAbeam 622-M OTU

The units came with a factory-supplied junction box. The inside of the box is shown in Figure 2-19. Payload fiber, management Ethernet, and power connections are made within this box. The two orange fibers to the right are the multi-mode fibers connecting the SmartBits to the OTU. Also visible on the lower fiber is the 10 dB attenuator inserted in the output line to enable the OTU and SmartBits to communicate. Figure 2-20 shows the junction box mounted on the support pole just below the OTU. The smaller utility box in this figure is the in-house fabricated power supply housing. It contains the 110 vac to -48 vdc power supply.

Table 2-3   SONAbeam 622-M Specifications

Manufacturer			fSONA Communication Corp 140-11120 Horseshoe Way Richmond, B.C. Canada
Model			SONAbeam 622-M
Cost			$40,522.69/pair w/training
Purchase Date			3/15/04
Data rate			OC-12 (622 Mbps)
Range Max		mi / m	3.4 / 5500
Transmitters		No.	4
Tx Type			Directly modulated laser diode (OOK)
Tx Wavelength		nm	1550
Tx Power		mW	560 (4 x 140)
Tx Beam Divergence (nominal)		mRad	2
Receiver Type			APD
Receiver Dia		in/cm	8 / 20
Auto-track			No
Interfaces Types		Fiber	Single mode or Multi-mode
Interface Connector		Data	SC
		Management	RJ-45 or DB9
Voltage			-48 vdc
BER			10^{-12}
Environment	Max Operating Temp		140 F
	Max Operating wind		100 mph
Laser Safety Class			1M (eye safe)
Assigned IP Address		#1	128.217.108.178
		#2	128.217.108.179
Assigned Subset Mask		#1 & #2	255.255.255.0
Assigned Gateway		#1 & #2	128.217.108.10
Serial Numbers		#1	1130050858
		#2	1130030756

Figure 2-19   **SONAbeam Junction Box Open**

Figure 2-20   **Pole Mount with OTU, Connection Box & Power Supply Box**

Bit errors were forced by choking down the receiver area of fSONA #1 using aperture masks that blocked a large portion of the 8-inch receiver aperture. Both fixed and variable area masks were used. Figure 2-21 shows a mask with a variable area aperture and two fixed apertures. The fixed apertures are taped over to make the variable units the only active area. A mask adapter was machined and attached to the receiver shroud on fSONA #1. This adapter allowed for the quick installation and positioning of the masks. Positioning was achieved by rotating the mask opening until the "sweet spot" was obtained. Aperture openings were machined off center since the internal geometry of the fSONA receivers blocks the center 1-inch diameter. In addition, the 12 and 6 o'clock positions are blocked by internal support struts. Multiple openings were machined in various masks, but only one opening was used at a time. The others were covered with tape. Tests were conducted with Aperture diameters of 8.0, 3.0, 2.0, 1.25 and 1.0 inch. Figure 2-22 is a close-up of the variable size aperture.

Figure 2-21   Aperture Adapter and Mask

Figure 2-22   Aperture Adapter and Mask

2.6.2 AirFiber 5800

The AirFiber FSO system uses single beam AirFiber 5800 OTUs, each with auto-track capability. The OTU is shown in Figure 2-23 with its fiberglass shroud removed. The auto-track feature and the built-in cameras enable both OTUs to quickly self-align themselves to provide an optimum link. The AirFiber system includes communication and feedback between the two OTU peers. During operation, the auto-track feature maintains optimum performance even with slight wind loads or building movements due to uneven diurnal or seasonal heating. The trailer-mounted AirFiber OTUs were tested only in the EDL parking lot at a distance of about 175 feet. The tests confirmed the AirFiber units were fully operational after two years of storage. Specifications for these units are presented in Table 2-4.

Figure 2-23   AirFiber 5800 With Fiberglass Shroud Removed

Table 2-4   AirFiber 5800 Specifications

Manufacturer			AirFiber San Diego, CA (Out of Business 2/26/03)
Model			5800-0623-MM
Cost			$23,724 /pair w/training
Purchase Date			12/12/02
Data rate			OC-12 (622 Mbps)
Range Max		mi / m	TBD
Transmitters		No.	1
Tx Type			TBD
Tx Wavelength		nm	785
Tx Power		mW	18
Tx Beam Divergence (nominal)		mRad	0.5
Receiver Type			APD
Receiver Dia		in/cm	3 / 7.5
Auto-track			Yes
Interfaces Types		Fiber	Single mode or Multi-mode
Interface Connector		Data	SC
		Management	RJ-45 or RS-232
Voltage			-48 vdc
BER			10^{-12}
Environment	Max Operating Temp		170 F
	Max Operating wind		120 mph
Laser Safety Class			1M (eye safe)
Assigned IP Address		#1	128.217.108.176
		#2	128.217.108.177
Assigned Subset Mask		#1 & #2	255.255.255.0
Assigned Gateway		#1 & #2	128.217.108.10
Serial Numbers		#1	00025000001017
		#2	00025000001035

2.6.3 SmartBits

An existing SmartBits, shown under the laptop in Figure 2-15, was used to verify the link quality for the fSONA units. The SmartBits created varying size data packets that were sent through the FSO communication link at OC-12 data rates. The SmartBits compared the data sent with the data received and produced a report on Throughput and Packet Loss.

The SmartBits was populated with a pair of ATM OC-12 driver cards in slot positions 17 and 19. For these tests, Card 19 was usually the transmitter and Card 17 was the receiver. A jumper fiber was also connected from the Card 19 Rx to the Card 17 Tx to complete the circuits. Figure 2-24 shows an earlier field test setup on Schwartz Rd.

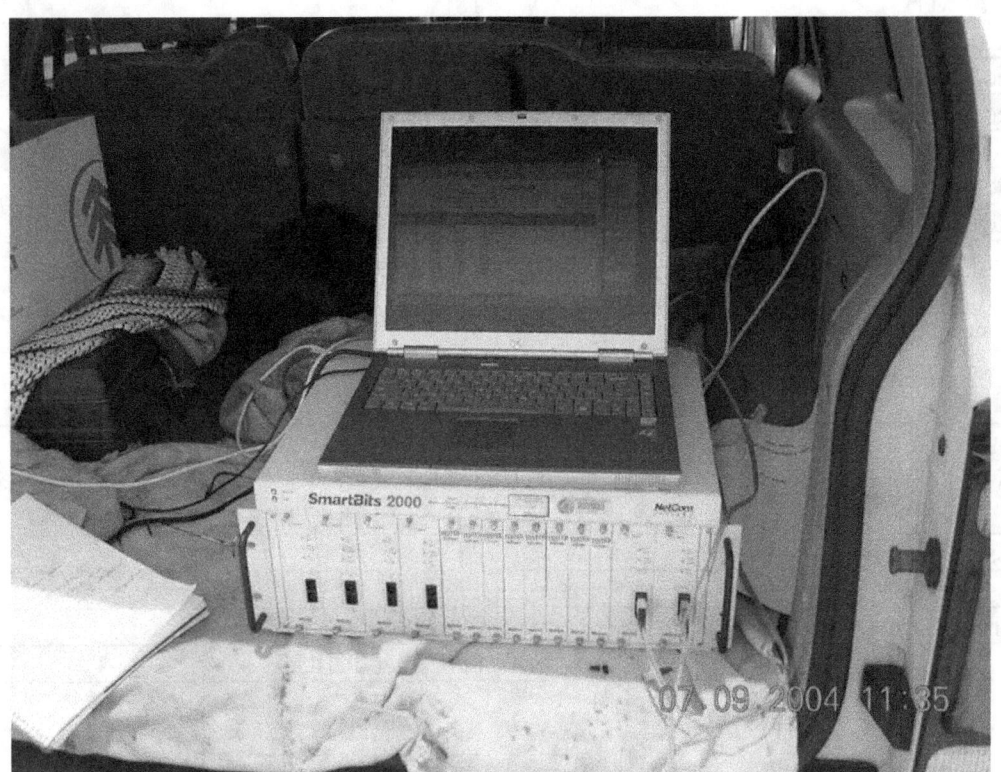

Figure 2-24   Field Testing with the SmartBits & a Laptop

2.6.4 Pattern Generators

Two different but interchangeable pattern generators were used during Scintillation testing. The primary unit was an Advantest D3186 10 Gbps pattern generator. The Anritsu MP1763C was also used for some tests. This equipment was provided and operated by FSI personnel.

2.6.5 **Error Detectors**

Two different but interchangeable error detectors were used during Scintillation testing. The primary unit was an Advantest D3286 10 Gbps error detector. The Anritsu MP1764A 10 Gbps unit was also used for some tests. This equipment was provided and operated by FSI personnel and located within the FSI trailer or van.

2.6.6 **Receiver**

An Agilent 83440C 20 GHz receiver was used downstream of fSONA #1 to convert the optical output for the fSONA to an electrical signal. This was necessary because both brands of error detectors required electrical inputs as opposed to fiber inputs. This equipment was provided and operated by FSI personnel and located within the FSI trailer or van.

2.6.7 **Clock and Data Recovery (CDR)**

A Mindspeed M21012 CDR was used to separate the clock and data signals and prepare them for error detector measurements. This equipment was provided and operated by FSI personnel and located within the FSI trailer or van.

2.6.8 **Weather Station**

Local weather data was measured using a Campbell Scientific weather station shown on the left in Figure 2-25. The weather station had its own control and data logging software that ran on a laptop. The station was later mounted on its own trailer. This equipment was provided and operated by FSI personnel.

2.6.9 **Scintillation**

A Scintec BLS900 scintillometer was used during SLF testing. The receive unit is shown in Figure 2-25 on the right of the weather station. The transmit section is shown in Figure 2-26. The two components were positioned 1 km apart, the same as the FSO units. This equipment was provided and operated by FSI personnel.

Figure 2-25   **Wx Station and Scintillometer Rx on SLF**

Figure 2-26   **Transmit Portion of Scintec Scintillometer**

2.6.10 SONAbeam Terminal Controller Software

The main Operating System for the fSONA 622-M is the Terminal Controller Software. An upgraded version was loaded prior to the Phase 4 tests. The software description provided in the Phase 3 report still applies.

2.6.11 SmartApplications

SmartApplications is the operating system software for the SmartBits. This software was described in the ECT Phase 3 report.

2.7 TEST RESULTS

FSO testing consisted of measurements at four test locations with two types of tests at each location. The four locations and the general test objectives at each are summarized in the following table.

Table 2-5 Summary of Test Locations & Objectives

Location	Test Objectives
SpaceHab	Laboratory measurements of fSONA output power and receiver configuration
EDL East Parking Lot	AirFiber OTU checkout, fSONA OTU checkout & procedure testing
Schwartz Road	Baseline testing and SLF dry run testing
SLF	Scintillation effect testing at 1 km

FSO testing was conducted from November 2004 through August 2005. Initial testing was delayed by recovery efforts from Hurricanes Charley and Frances. During the test period, the fSONA 622-M units worked without any noticeable problems. The most difficult part of operating the OTUs was establishing the initial FSO link alignment and link performance. The AirFiber 5800 OTUs were found to be fully operational.

2.7.1 Prior Art and Historical Rationale for Testing Methodology

Recent research on the effects of optical scintillation on the probability of error, or bit error-rate (BER), involved the extension of the Rytov approximation into the strong fluctuation regime[5] and the use of a (spatially) partially-coherent source.[6] These advances in the theory enable addressing optical scintillation degradation concerns for evaluating a wide range of future lasercom needs, including Earth-to-Mars, Earth-to-orbit, and Earth-to-Moon FSO links.

For addressing the theoretical limits that exist for another set of future FSO needs, such as for achieving higher efficiency lasercom systems for terrestrial point-to-point links, enabling communication through rocket exhausts with FSO systems, developing lowered probability of intercept covert communication systems, and enhancing military laser target designator weapons systems, more complex theories of optical scintillation, for predicting both amplitude and phase variations, are needed. After all, as documented in prior NASA research at JPL, space-to-ground links for satellite lasercom applications involve beam propagation through less air mass than horizontal terrestrial path links exceeding 10 km in length.[7] The terrestrial point-to-point FSO links operating over longer distances are more difficult links to close than the more esoteric longer range and more glamorous Earth-to-Mars, orbit, and Moon links.

Classical approaches to studying optical wave propagation are based primarily on studying simplified uniform plane wave and spherical wave models. Plane wave models, for example, are often used in describing the properties of starlight or laser beams passing through the Earth's atmosphere from space. Zernike's phase contrast method of observation involves assuming the existence of a *phase object* which alters the phase but not the amplitude of an incident plane wave.[8] Unfortunately, such simplified approaches do not account for the various effects that are caused by the finite size of a beam wave and its diverging and focusing capabilities.

The traditional method to address and overcome the limitations of simplified modeling approaches is to develop theory for a basic wave model based on the lowest order Gaussian-beam wave, which is characteristic of a single transverse electromagnetic wave (TEM_{00}). This approach has the advantage of developing optical scintillation theory that provides, as limiting cases, the same results as have been previously shown by the earlier classical approaches. It also permits enhancing one's ability to address many of the first-level concerns that result from assuming a random media such as the atmosphere, for which small index-of-refraction fluctuations induced by random temperature variations are the primary concern.[9]

---

[5] L. C. Andrews, R. L. Phillips, and C. Y. Hopen, *Laser Beam Scintillation with Applications* (SPIE Press, Bellingham, WA (2001).

[6] O. Korotkova, L Andrews, R. Phillips, "Model for a partially coherent Gaussian beam in atmospheric turbulence with application in Lasercom," Opt. Eng. 43(2) 330-341 (Feb. 2004).

[7] A. Biswas, M. Wright, B. Sanii, N. Page, "45 Km horizontal path optical link demonstrations," Proc. SPIE 4272, 60-71 (2001).

[8] M. Born, E. Wolf, *Principles of Optics*, Pergamon Press, New York, NY (1975), Fifth Edition, pp. 424-428.

[9] L. C. Andrews, R. L. Phillips, *Laser Beam Propagation through Random Media*, SPIE Press, Bellingham, WA (2001).

Optical scintillation theory developed in this way is entirely consistent with Born and Rytov weak fluctuation theories, and permits developing spectral representations for the first-order and second-order complex phase perturbations that are consistent with the Rytov approximation.[10] Building off an extended version of the Rytov approximation is a method that characterizes optical scintillation into the focusing and saturation regimes. This extension of the conventional Rytov theory is based on the notion that only scale sizes smaller than the spatial coherence radius or larger than the scattering disk contribute to optical scintillation under moderate-to-strong irradiance fluctuations. The theory has also been extended to account for large receiver apertures that lead to a reduction in scintillation known as the averaging effect.

For the measurements made at the SLF to validate these previously unvalidated theories, two commercially-available instruments built by fSONA were used for the optical transmitter and receiver to perform the experiments. The purpose of the experiment was to measure the effect of aperture averaging on the BER and additionally to compare these measurements with the published theory. Thus, various aperture sizes were used at the receiver end of the propagation path to study the decrease (improvement) in BER as the size of the receiver aperture was increased. Because the theory of optical scintillation in moderate-to-strong fluctuation regimes is relatively new, it has not been satisfactorily validated through the collection of experimental data. Consequently, the taking of experimental data is key to determining the limitations of this optical scintillation theory, and to validate its use in future *a priori* communication link planning activities in and around Kennedy Space Center.

2.7.2 **Testing Methodology**

The basic characteristics of the commercial instruments used during the experiments were described in Table 2-3. The nominal free-space diameter of the spot size in the receive plane of the lasercom beam from the laser beam divergence was 2 meters, typical for a 1 km spacing between transmitter and receiver. Even under fairly strong turbulence conditions (i.e., $C_n^2 \sim 5 \times 10^{-13} \text{m}^{-2/3}$), the beam diameter was only a few centimeters larger than 2 meters. Artificially reducing the receiver aperture from the 20-cm (8-in nominal) effective clear aperture, by controlling aperture mask diameters at the receiver, allowed simultaneously reducing receiver power levels while affecting signal to noise ratios within the data rate bandwidth. This permitted studying the effects of aperture averaging of scintillation effects for larger aperture masks on BER, at power levels for which moderate numbers of errors would occur during a reasonable time.

In addition to the normal factory-standard hardware of the fSONA SONAbeam 622-M, we sometimes added external neutral-density filters to the transmitting lasers. These transmitting laser add-ons enabled us to achieve more precisely calibrated control of the

---

[10] L. C. Andrews, R. L. Phillips, *Laser Beam Propagation through Random Media*, SPIE Press, Bellingham, WA (2001), Chapter 5.

transmitting laser power, as needed to achieve the desired signal-to-noise ratios in the optical receiver as required to study bit error rates. Although reducing the aperture reduces the signal power and external noise power equally, which clearly has no impact on the signal-to-noise ratio, the internal noise power is not reduced by reducing the aperture mask diameter. Hence, reducing the aperture diameter can be used to vary the signal-to-noise ratio due to the presence of internal noise.

2.7.3 **Instrument Operation**

Atmospheric conditions were continuously monitored during the experiments. Average values of refractive-index structure parameter, C_n^2, were provided by a Scintec BLS900 scintillometer over the time period of data collection. In particular, measured values of C_n^2 that were averaged over one minute periods are shown in Figure 2-27. Typical values were in the range of 10^{-14} to more than 10^{-13} m$^{-2/3}$. This same scintillometer instrument also provided the average cross wind speed during the time period of each experiment. Other environmental data instrumentation included a weather station that gave wind speed and direction, temperature, humidity, visibility, cloud ceiling, solar irradiance, atmospheric pressure, and runway surface temperature.

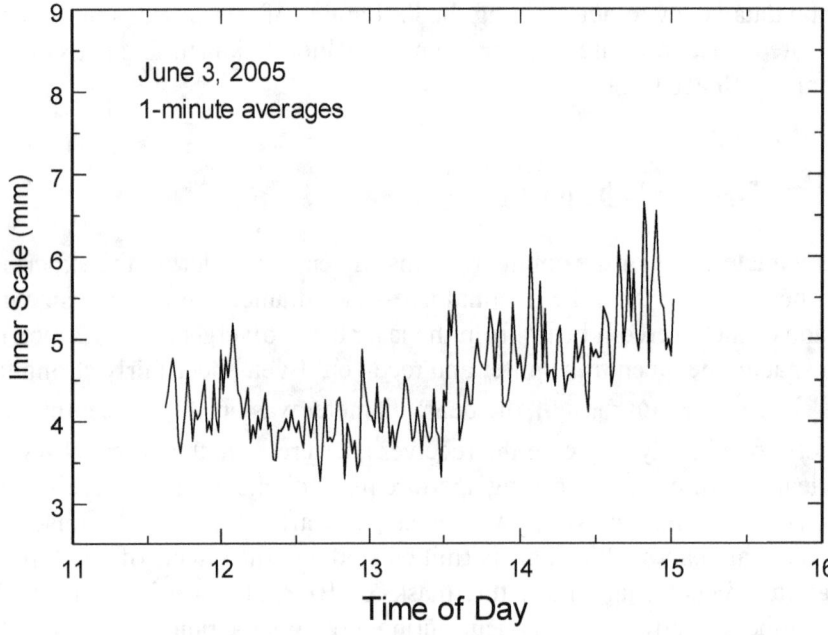

Figure 2-27   **Measured Average C_n^2 Values**

The outer scale of turbulence L_0 was estimated to be 1 m, equal roughly to half the height of the optical wave path above ground. Values of the inner scale of turbulence l_0 were inferred from the surface roughness of the Shuttle Runway and average wind speed, i.e.

$l_0 \sim const./V^{3/4}$, where V denotes the average wind speed. One minute average values of the wind speed produced the inner scale values shown in Figure 2-28. Typical values of the inner scale calculated by this method were between 3 and 7 mm.

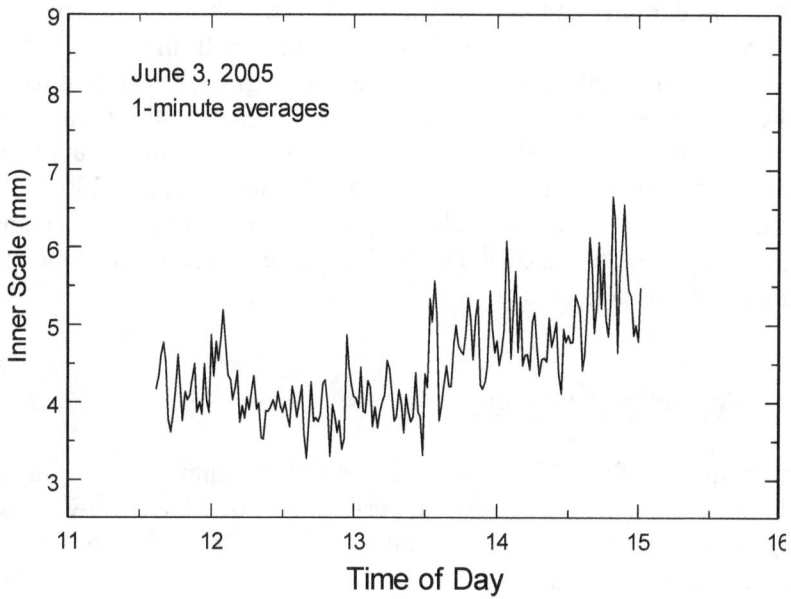

Figure 2-28   Values of Inner Scale from 1-min. Avg. Wind Speed

In addition to environmental data instrumentation, bit error rate data collecting instrumentation was also used during the experiments. An OC-12 SONET rate signal at 622.08Mbps was generated using an Advantest D3186 Pattern Generator (PG). The PG was configured to generate an NRZ-coded pseudo-random bit sequence (PRBS) 2^23-1 pattern data stream with a ½ mark ratio. There was no framing present on this signal. This PRBS signal drove a custom electro-absorption modulated laser (EML). This modulated laser signal was then attenuated to a level that matched the acceptable operating input optical levels of the fSONA optical transceiver. Within the fSONA transceiver, this optical signal was regenerated and then used to drive any or all of the four fSONA internal 1550nm Fabry-Perot laser transmitters. Throughout the majority of our experiments, only one of the fSONA transmitters was configured to be active at any one time. Prior to running SLF range experiments, this particular transmitter was first thoroughly characterized in a controlled optical lab bench environment using various calibrated instruments. We experimentally measured laser transmitter properties including linewidth, output power level settings, beam diameter, stability, extinction ratio, and eye diagrams, all of which combined to help us better understand the limitations of the hardware and to determine which optical transmitter power levels would be best suited for use during experiments.

The transmitted signal from the fSONA transmitter terminal was sent through free space (1 km) and entered the fSONA receiver terminal. The receiver terminal focused the collected 1550nm light onto an Avalanche Photo Diode (APD). This received optical signal was then regenerated inside the fSONA terminal using a trans-impedance amplifier (TIA), pre-amplifier, bandpass filter, post-amplifier, and limiter. The recovered data stream was then used to drive another internal laser to transport the received data out of the fSONA receiver terminal. Once the signal was removed from the fSONA receiver terminal, we needed to sample it again to recover the original transmitted pattern from our PG on the transmitter side. This was done using a high-sensitivity, multi-mode, 2.5Gbps multi-rate receiver module. This module drove a clock and data recovery (CDR) circuit which then fed the recovered data and clock to the Anritsu D3286 Error Detector. Automated instrument drivers controlled and synchronized data collection among the scintillometer, weather station and the fSONA's received optical power meter during the entire time that data was collected.

2.7.4 Theoretical Equations

The modulation format of the SONAbeam 622-M FSO equipment used in this study is considered to be bounded by the theoretical performance of OOK (ON-OFF Keying), despite being NRZ-coded (non-return-to-zero coded). This modulation type is chosen because the receiver does not take advantage of coherent methods of detecting transmitted NRZ coded signals, but instead uses only non-coherent OOK detection techniques. The significance of this is that the theoretical performance of the FSO equipment used is 3 dB worse than if coherent NRZ decoding were used.

Probability of error, which determines the theoretical Bit Error Rate (BER) curve performance, is given for NRZ-coded electrical data by[11]

$$\Pr(E) = \tfrac{1}{2} \cdot \text{erfc}\left(\sqrt{z}\right) \qquad (1)$$

where erfc(x) denotes the complementary error function and z is the data signal to noise power ratio. This same probability of error equation applies equally for any antipodal-encoded modulation (e.g., NRZ, BPSK, PRK, etc.) in which the cross-correlation, R_{12}, equals -1 between the two signaling states.[12] It does not, however, express the predicted probability of error for OOK (ON-OFF Keying.) For OOK (which is orthogonal-encoded modulation), the cross-correlation between the two signaling states equals zero, and this shifts the theoretically-predicted BER curve 3-dB worse than antipodal signaling for the

---

[11] R. Ziemer, W. Tranter, *Principles of Communications, Systems, Modulation, and Noise,* Houghton Mifflin Co., Boston, 1976, pp. 315-316.
[12] R. Ziemer, W. Tranter, *Principles of Communications, Systems, Modulation, and Noise,* Houghton Mifflin Co., Boston, 1976, pp. 319-320.

required signal-to-noise ratio that is necessary for achieving any particular desired BER performance.[13]

OOK is also sometimes called Intensity Modulated Direct Detection (IM/DD). In practice, true OOK is difficult to achieve for FSO systems, which are limited by the extinction ratios existing between the two states. OFF is never truly OFF. For the equipment used in this experiment, an extinction ratio of approximately 10 dB is all that is achieved for the 'OOK modulation' encoded states by the hardware. Despite being 'OOK-decoded', the effective modulation encoding is therefore actually NRZ-encoded for the hardware used in this experiment. The theoretical BER limits pertaining to NRZ rather than to OOK are most applicable for predicting absolute theoretical BER performance curve limits. However, due to non-coherent receiver operation, the theoretical BER limits for OOK (IM/DD) remain more representative for predicting the measured performance of the hardware under investigation. Hence, a choice of comparing against the theoretical BER curve pertaining to OOK (IM/DD) is taken as the appropriate theoretical limit for the present investigation, despite giving up 3-dB from the actual theoretical limit that also exists.

OOK additionally provides by far the better BER analytical model than NRZ for modeling the SONAbeam 622-M gear's BER performance, due to its assumption of full bit-window data bits, and non-coherent data detection in the actual hardware.

Probability of error for OOK-coded electrical data is given by:[14]

$$P_E = \tfrac{1}{2} \cdot \text{erfc}\left(\sqrt{\tfrac{z}{2}}\right) \qquad (2)$$

A theoretical 10^{-9} BER is achieved for approximately a 15.6 dB data signal to noise power ratio. Probability of error for OOK-coded optical data, detected with a photodiode (see Appendix B), is given by:

$$P_E = \frac{1}{2} erfc\left(\frac{1}{2}\sqrt{\frac{z}{2}}\right) \qquad (3)$$

A theoretical 10^{-9} BER is achieved for approximately a 21.6 dB signal to noise power ratio. This BER performance level applies only in the absence of atmospheric turbulence, which only increases the effective implementation loss of hardware measured against theoretical OOK-coded BER performance curves. Such OOK-coded or IM/DD data has

---

[13] R. Ziemer, W. Tranter, *Principles of Communications, Systems, Modulation, and Noise,* Houghton Mifflin Co., Boston, 1976, p. 315.

[14] IRIG Standard 106-04, Part I, Appendix C, p. C-1 (p. 356 of 495.) http://www.ntia.doc.gov/osmhome/106-04.pdf , approved for public release, unlimited distribution.

also been known as unipolar Non Return to Zero Level, NRZ-Level, and NRZ Change encoded data when discussed in IRIG telemetry standards.[15]

The data signal to noise power ratio in either case (i.e., for either NRZ or OOK) is z which is defined as

$$z = \frac{S \cdot \tau}{N_0} \tag{4}$$

where S is the total signal power in Watts, τ is the bit period in seconds, and No is the noise power spectral density in Watts/Hz.

The inverse of the bit period is the bit rate bandwidth. That is,

$$\frac{1}{\tau} = R \tag{5}$$

Re-writing Eqn. 4, the physical interpretation becomes the ratio of the total signal power to the total noise power, i.e., the signal to noise ratio, *as measured in the bit rate bandwidth*. That is to say:

$$z = \frac{S}{N_0 \cdot R} = \frac{E_b}{N_o} \tag{6}$$

Provided that average signal-to-noise ratio, denoted by $<SNR>$, is defined in exactly the bit rate bandwidth, $<Eb/No>$ becomes exactly equal to the $<SNR>$. The key point is that the signal to noise ratio is *not* what determines the BER performance, *unless* proper assumptions are made relative to the bandwidth over which the noise power spectral density is assumed to be integrated.

In the presence of atmospheric turbulence, the probability of error (BER) is considered to be a conditional probability that must be averaged over the probability density function (PDF) of a random signal s to determine the unconditional BER. This action leads to the expression:

$$\Pr(E) = \frac{1}{2} \int_0^\infty p_I(s) \operatorname{erfc}\left(\frac{s}{2} \sqrt{\frac{<SNR>}{2}} \right) ds \tag{7}$$

where $<SNR>$ is the average signal-to-noise ratio and $p_I(s)$ is the PDF of the normalized signal (unit mean).

---

[15] M. A. Al-Habash, L. C. Andrews, and R. L. Phillips, "Mathematical model for the irradiance PDF of a laser beam propagating through turbulent media," Opt. Eng. 40, 1554-1562 (2001).

Rather than using the traditional log-normal PDF, which is valid only under assumptions of weak turbulence, we find that the gamma-gamma PDF is more accurate under both weak and strong turbulence conditions. Hence, in the analysis, we use the gamma-gamma PDF defined by[13]:

$$p_I(s) = \frac{2(\alpha\beta)^{(\alpha+\beta)/2}}{\Gamma(\alpha)\Gamma(\beta)} s^{(\alpha+\beta)/2-1} K_{\alpha-\beta}\left(2\sqrt{\alpha\beta s}\right), s > 0 \qquad (8)$$

where parameters α and β are related to the reciprocals of the large-scale and small-scale irradiance fluctuations, respectively, of the random signal. In particular, the scintillation index of the irradiance fluctuations takes the form:

$$\sigma_I^2 = \exp\left[\sigma_{\ln x}^2(l_0, L_0, D) + \sigma_{\ln y}^2(l_0, D)\right] - 1, \qquad (9)$$

where D is the receiver aperture diameter, and where we define:

$$\alpha = \frac{1}{\exp\left[\sigma_{\ln x}^2(l_0, L_0, D)\right] - 1}, \quad \beta = \frac{1}{\exp\left[\sigma_{\ln y}^2(l_0, D)\right] - 1}. \qquad (10)$$

Expressions for the large-scale and small-scale irradiance fluctuations are from published equations.[16]

The implementation loss measured relative to the NRZ theoretical curve is what determines the absolute quality of any particular set of hardware, as well as the true performance impact introduced by an assortment of degradations ranging from perhaps the timing uncertainty in particular bit synchronizers to, as is the topic of interest in this paper, the BER effects of optical scintillation versus aperture diameter size. For applying the results to real hardware, however, while accounting for real hardware limitations, the OOK theoretical curve is often more applicable, such as is the case for the measurements reported herein.

2.7.5 Detailed Results

Measured and theoretical values of the BER for a receiver aperture of 1.25 inches are shown in Figure 2-29 for one-minute averages. The rather meaningless scatter-diagram of Figure 2-29 shows the need for increasing the averaging time in order to make

---

[16] L. C. Andrews, R. L. Phillips, and C. Y. Hopen, *Laser Beam Scintillation with Applications* (SPIE Press, Bellingham, WA (2001).

meaningful comparisons between theory and measured data. We therefore performed averages of all quantities over a five-minute period and plotted the results in Figure 2-30. In Figure 2-30, we also included measured data over the time period from 1:45 PM to 3:00 PM with corresponding predicted theoretical values. During this latter time period, however, we often lost lock on the BER instrumentation and most of the measured data was at either high BER values or nearly zero. We believe taking data over longer time periods (5-10 min.) would eliminate most of the remaining erratic behavior observed in comparing theoretical predictions with measured data.

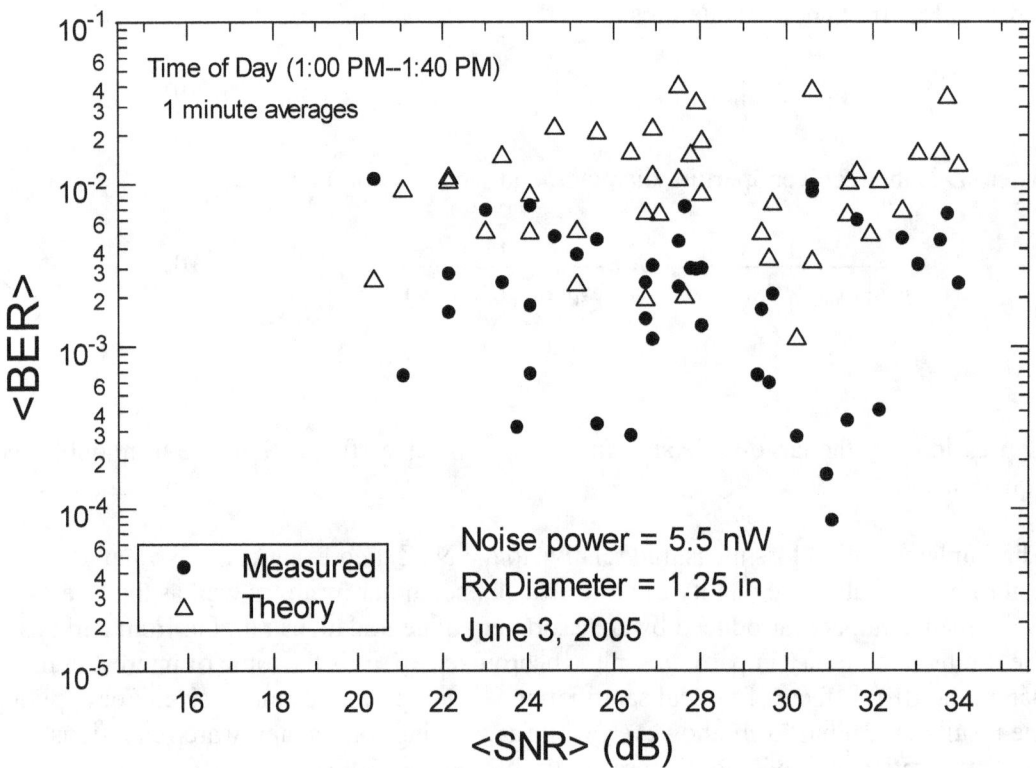

Figure 2-29 1-min. Avg. BER Vs. SNR

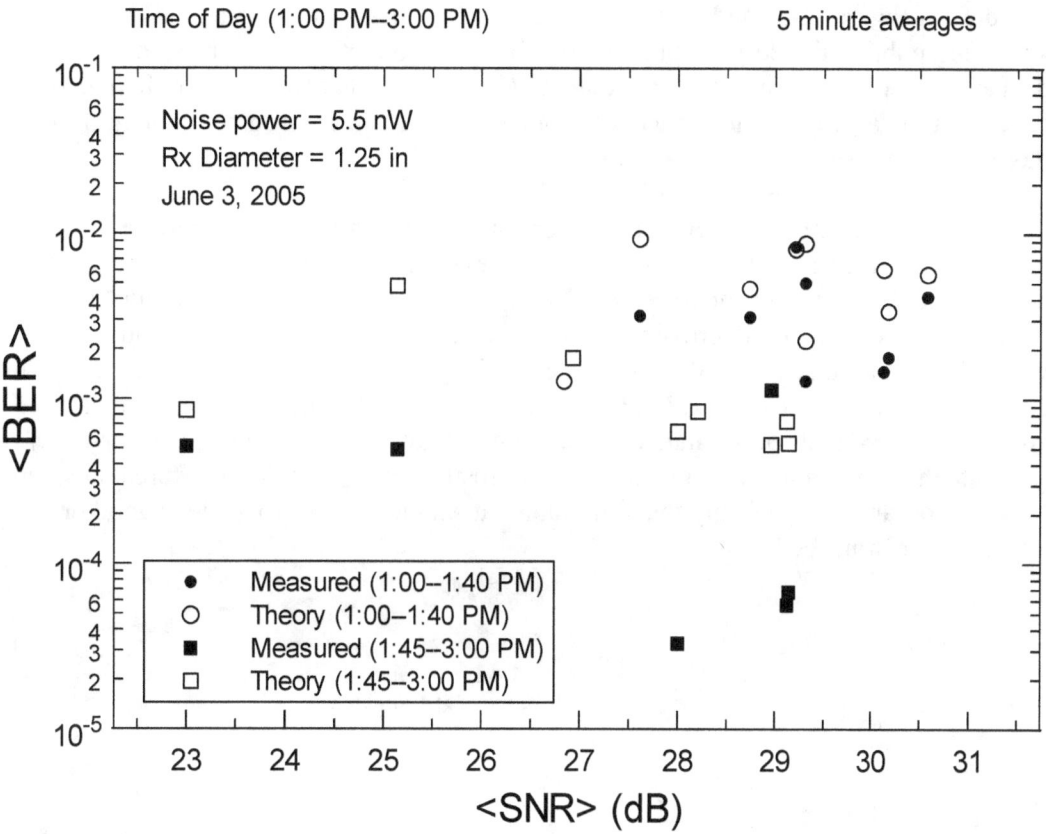

Figure 2-30   5-min. Avg. BER Vs. SNR

For both Figures 2-29 and 2-30, to compare theoretical predicted results with these measured quantities, it was necessary to first calculate the mean noise power at the receiver. This back-to-back measurement was done under laboratory conditions in the absence of atmospheric effects and found to be 5.5 nW for Transmitter Power Level 7 (highest level). Transmitter Power Level 7 for the fSONA transmitter was chosen specifically since it had the best extinction ratio (best eye diagram) in contrast to Power Levels 1 through 6, thereby reducing the introduction of hardware effects in accurately measuring the BER performance impacts of atmospheric turbulence. Using this value for the mean noise power, and the measured average signal power at the receiver, the resulting theoretical BER was calculated using Eq. (7) and also shown in Figures 2-29 and 2-30. The parameters (9) of the gamma-gamma distribution (8) were based on path-averaged values of C_n^2, calculated inner scale values over the period of the experiment, and the strong fluctuation scintillation theory.[17]

---

[17] L. C. Andrews, R. L. Phillips, and C. Y. Hopen, *Laser Beam Scintillation with Applications* (SPIE Press, Bellingham, WA (2001).

The data of Figures 2-29 and 2-30 show that we are still at a preliminary stage in validating published theory for predictive purposes, and that considerably more collected data are required to complete the validation. Considering that BER is a predictive tool for estimating average performance in an otherwise random environment, the need for more data is not surprising.

Figure 2-31 shows measured BER and error count data versus receive aperture diameters of 2-inches, 3-inches, and 8-inches (full-aperture) over a 1 km link. For aperture diameters of 2-inches, the measured BER is seen to have often clustered around 10^{-3}, showing only occasional improvements to 10^{-7} to 10^{-10}. The full aperture (8-inches), when checked around 1:55 PM, provided essentially 10^{-12} or better (error-free) BER with only a few instances of 10^{-5} BER. However, for receive apertures of 3-inches measured around 3:00 PM, BER performance ranges from 10^{-9} to 10^{-12} or better (i.e., error-free.) The data therefore indicate that over a 1 km terrestrial link, a receive aperture diameter of 3-inches, or larger, is probably the minimum that should be used to achieve good or adequate performance.

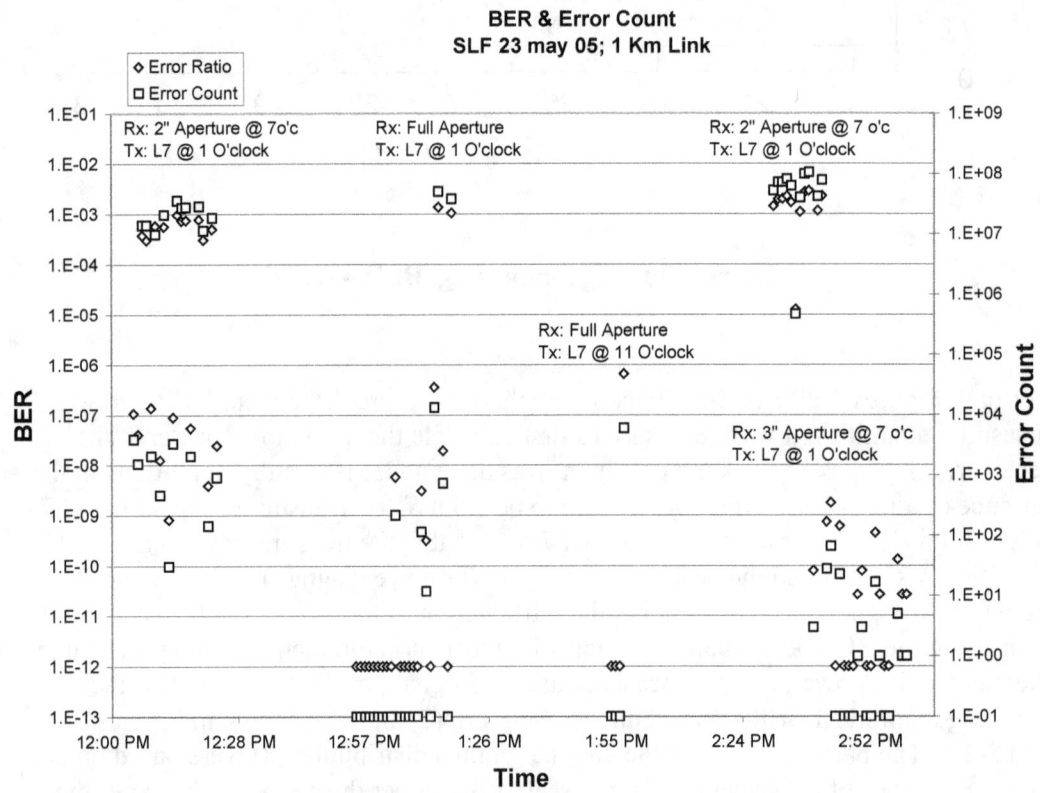

Figure 2-31   BER & Error Count Vs. Time of Day & Aperature Dia.

Although the 3-inch minimum diameter for achieving good BER performance over a 1 km link appears to be the minimum diameter that should be employed, there are times when a smaller diameter aperture can nonetheless provide surprisingly good performance, lasting through appreciable periods of 'good seeing.' Figure 2-32 shows measured BER data for 1.0-inch and 1.25 inch apertures, respectively, measured over a 1 km long link. For the 1-inch diameter receive aperture, a BER of around 10^{-2} to 10^{-3} is as good as could be achieved. Yet, for a 1.25-inch diameter receive aperture, BER performance often achieved 10^{-6} to 10^{-12} to error free. These data also show that increasing aperture averaging through increasing receive aperture diameters only slightly can be a major factor in determining, or at the very least improving rather dramatically, the BER performance of a free-space optical link. Aperture averaging can clearly work to improve measured BER performance dramatically for only very slight increases in receive aperture diameter under appropriate atmospheric turbulence conditions.

The variations of the refractive-index structure parameter, C_n^2, must also be considered, in addition to the variations in the receive aperture diameter for completeness, less the importance of only slight differences in receiver aperture diameter become overstated. Figure 2-33 shows C_n^2 variation for June 3rd, applicable for the measured data shown in Figures 2-27 through 2-30 & 2-32 vs. measured BER. As one would expect, the peak of C_n^2 occurs around local solar noon, perhaps delayed slightly (10-15 minutes or so) at which time the scintillation is near its peak diurnal value. It then decreases rather rapidly for approximately an hour, as the sun descends toward the western horizon, before plateauing to nearly a constant value.

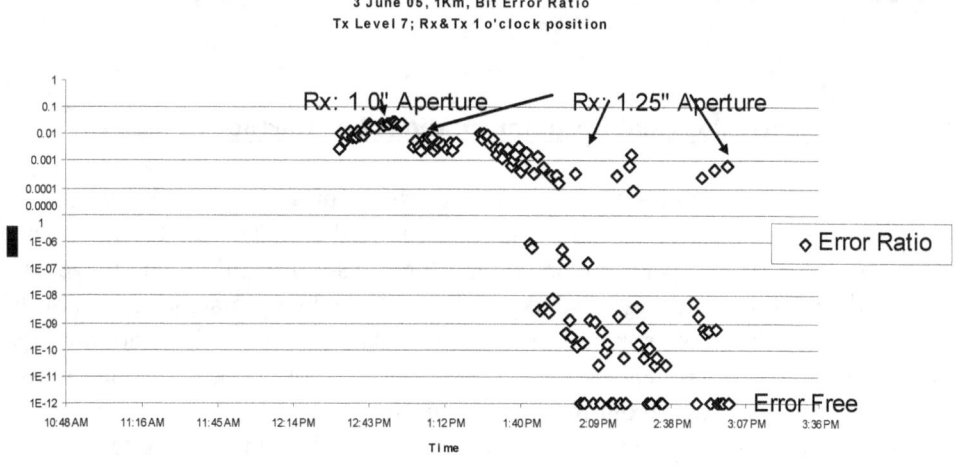

Figure 2-32   BER Vs. Time of Day @ 1.0 & 1.25 Aperture Dia.

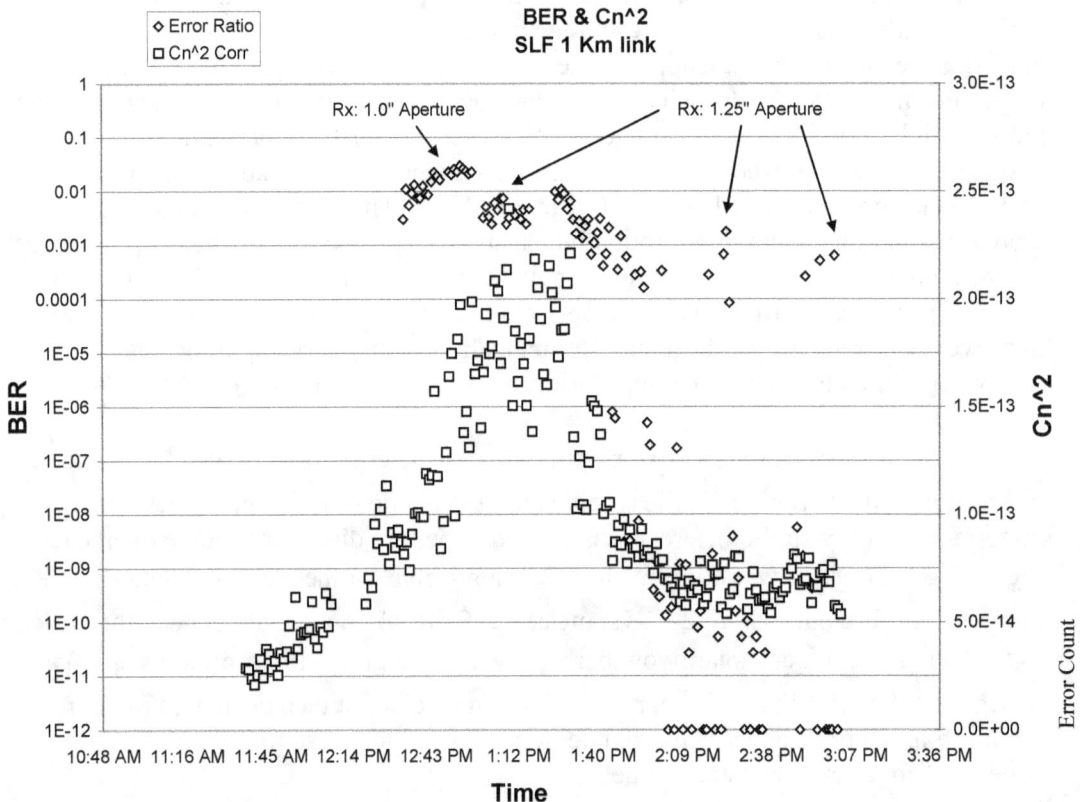

Figure 2-33 BER & C_n^2 Values Vs. Time of Day @ 1.0 & 1.25 Aperture Dia.

2.7.6 Conclusions and Recommendations For Further Research

Results of FSO scintillation tests conducted at the SLF runway at KSC have been presented. The absolute accuracy of the theoretical predictions of BER performance versus measured BER for specific receiver aperture masks are still inconclusive based on the limited data obtained during these experiments. The theory matches much of the measured data quite well but some of the measured data does not fall where the theory predicts. Additional experiments are necessary to make more definitive statements about the subtle differences between our theory and measured results. Longer term data collection would enable improving the accuracy of the theory through better understanding the limitations of the presently-developed predictive equations. Nonetheless, the use of the theory as published appears adequate to enable *a priori* performance estimation of FSO link performance within the unique KSC environment.

The fundamental characteristic of improving BER through increasing receive aperture diameter, though, is clear. Differences in aperture averaging through increasing receiver aperture diameters only slightly can sometimes greatly improve measured BER performance. Additionally, there are rather long moments of 'good seeing' during which times surprisingly good BER link performance can be achieved despite the use of rather small receiver aperture diameters.

More research is needed to understand better the limitations and benefits of aperture averaging to achieve better BER performance for terrestrial point-to-point data links.

The minimum aperture size that should be considered for implementing very short FSO links at KSC is 10 cm, and aperture diameters of 32 cm, or larger, should be considered as being the minimum practical diameters that should be deployed over paths of 1 km or longer within the hot Florida locale of KSC.

In addition to the receiver aperture averaging advantage reported in this paper, transmitter aperture averaging effects also need to be investigated in the course of future research, time permitting.

ECT – Phase 4

3.0

EXTENDED RANGE WIFI

3.1 BACKGROUND

Operation of Extended-Range (E-R) Wi-Fi equipment can be accomplished using licensed or unlicensed frequency bands. Acquiring a license to operate such equipment can be time-consuming and costly. After considering the project's budget and timeline, it was decided to utilize an unlicensed frequency band, to expedite testing, while following NASA Spectrum Management rules. Part-15 authorized unlicensed equipment could be tested without the need for NTIA Federal Frequency approvals, and was the choice made to expedite testing.

Hence, our E-R Wi-Fi testing under the ECT project was conducted around the DragonWave AirPair 100 with equipment operating at the unlicensed 24 GHz ISM (Industrial, Scientific, Medical) FCC-allocated frequency band. The DragonWave system was the first and only system available that would operate at the selected frequency band at the time that this equipment was procured, although additional vendors are scheduled to have started releasing competing equipment subsequent to the start of our testing.

3.2 BASIC E-R WI-FI THEORY

Wi-Fi (Wireless Fidelity, i.e., wireless Ethernet) is intended for use over short communication ranges and was investigated and its measured performance was discussed previously in the prior years' final reports. It is not the focus of activity in this year's report.

Nonetheless, Wi-Fi provides excellent wireless Internet connectivity over distances up to a few hundred meters for laptops and related assorted wireless networked appliances (e.g., shared printers, print servers, cameras, server hard-drive storage banks, speakers, scanners, etc.) Although certainly useful over short ranges, Wi-Fi does not provide connectivity over the longer ranges needed for implementing the future Range vision of First Mile/Last Mile connectivity over longer, or extended-range distances. Subsequently, the focus for this year was on achieving longer range communication, with essentially a Wi-Fi compatible wireless link, through utilizing Extended-Range (E-R) Wi-Fi.

Most E-R Wi-Fi communication equipment operates fundamentally through providing an Ethernet interface at two points, with an interconnecting microwave or millimeter wave communication wireless or radio frequency link being used for providing virtual wired connectivity between the two points. Hence, existing E-R Wi-Fi tends to be point-to-point radio links, although they are links that behave much like shorter-distance wired local area networks (LANs) implemented with traditional Wi-Fi.

For such E-R Wi-Fi systems, microwave high-gain dish antennas are used, typically with beamwidths of less than 5 degrees, with higher-gain antennas of course having smaller beamwidths. Transmit powers of a few milliwatts up to hundreds of milliwatts are used by authorized unlicensed systems, with higher transmit powers of Watts or tens of Watts being reserved for E-R Wi-Fi systems requiring FCC or NTIA licensing.

The advantage of such systems, whether permitted as licensed or unlicensed, is that effective isotropic radiated power (EIRP) is increased through using the high-gain dish antennas, and frequency re-use is likewise possible within a general geographic area through taking advantage of sectoral or otherwise geometry-restricting antenna radiation patterns. The use of common high-gain antennas, used for both transmit and receive functions, works to increase received signal levels, while likewise also reducing the amount of transmitter power required to achieve high fidelity low Bit Error Rate (BER) data links.

E-R Wi-Fi systems are not the only way to provide longer range Wi-Fi compatible operation. Alternative systems are under development by industry leaders as well, for providing wide area network (WAN) Ethernet communications, instead of only providing point-to-point communication links, such as used for E-R Wi-Fi. These systems include EvDO (Evolution- Data Only) and Wi-Max, typically used for providing wireless Internet connectivity over wide areas of major metropolitan areas. Wi-Max will be investigated next year.

3.3 CHOICE OF E-R WI-FI FOR TESTING

Initially, an E-R Wi-Fi system sold by Mobicomm in the Netherlands was investigated. It was available in a variety of configurations, operating on one of a range of existing frequency allocations usually reserved for PCS or cell phone licensed operation within the United States. However, there were two major issues with procuring and testing this system. First, Mobicomm ceased production of their first designs, and entered a re-design activity, to overcome design deficiencies in their initially sold designs, prior to the receipt of project funding that could enable the procurement of an E-R Wi-Fi system for testing. Second, through virtue of operating on FCC-authorized frequencies reserved for cell phone transmissions within the US, obtaining permission from the NASA Spectrum Manager, and from NTIA, while also coordinating with the FCC, appeared to be a largely impossible task, at least within the time constraints of desiring to test an E-R Wi-Fi system this fiscal year.

Fortunately, the FCC authorized a new unlicensed ISM band at 24 GHz for the purpose of providing spectrum for achieving E-R Wi-Fi operation within the US. DragonWave, a small Canadian wireless hardware company provider outside of Ottawa, responded through quickly re-designing and releasing a modified version of their system requiring FCC licensing but which could operate on the newly authorized 24 GHz unlicensed ISM band. This was the system that was procured for testing.

3.4 TEST DESCRIPTION

Initial tests were to become familiar with the system and to evaluate its performance at a very short distance (approximately 60 meters). Later tests were to evaluate performance at increasing link distances (1, 1.5, 2.0, and 2.5 miles). Weather testing was not accomplished in this phase.

3.5 TEST OBJECTIVES

The Extended Rage Wi-Fi test objectives were as follows:

- Evaluate COTS Extended-Range Wi-Fi equipment for possible future use at KSC
- Identify any fundamental shortcomings that must be mitigated in commercial Extended-Range Wi-Fi communication technologies prior to integrating this technology into future range architectures.

3.6 TEST SETUP

The test setup utilized when evaluating the Dragonwave E-R Wi-Fi AirPair 100 equipment was similar in all test cases. The primary difference between tests was just the distance between the two ends of the system. Initially, the system was tested at a very short distance (approximately 60 meters). This was done primarily to become familiar with the hardware, the software, and to practice aligning the antennas, while doing initial setup testing in the EDL back parking lot. The system was then moved to Schwartz Road to perform tests over greater distances.

The basic setup was to place the two AirPair 100 units a fixed distance apart and establish a link. Initial alignment was usually accomplished using a 12X power rifle scope borrowed from the FSO testing described previously, attached to a frame housing that was modified to permit attachment of the scope in addition to providing a mounting bracket for the RF head of the AirPair 100 unit.

Once an initial link was established, the pointing of link antennas was refined and optimized using the Web Interface on a laptop computer to provide feedback of the received signal level reading of the AirPair 100 system. Fine adjustment was accomplished using jacking screws on both azimuth and elevation lever arms.

WSTTCP (a Windows socket port of the BSD TTCP, i.e., of the Unix BSD Test TCP, software) was used to test data traffic throughput performance over the wireless link. This software runs on the laptops at each end of the wireless network. One laptop was configured to transmit test data and the other to receive. The receiver reports the results of the test back to the transmitter for completion purposes via the Ethernet connection. Tests were executed using two different data buffer sizes. The results of each test reported the effective data throughput provided by the wireless network enabled by the AirPair 100 system.

The primary independent variable for all tests was link distance. A summary of test locations and distances is included in the following table. Specific details about each test location are included in the following sections. The AirPair 100 system, test equipment, and software are described in later sections.

Table 3-1   Test Locations

No.	Location	One-Way Distance
1	Schwartz Road	1.0 mile
2	Schwartz Road	1.5 miles
3	Schwartz Road	2.0 miles
4	Schwartz Road	2.5 miles
5	Shuttle Landing Facility	3.2 miles

3.6.1   Schwartz Road Test Setup

Schwartz Road is a remote East-West road at KSC that runs relatively straight for 2.7 miles. Most of the testing was performed at this location. Trailers were positioned off to the side of the road at various separation distances. Initial testing was performed at 1.0 mile separation between trailers. This was followed by tests at 1.5, 2.0, and 2.5 miles. The west trailer remained fixed during all tests.

Trailer #2 was positioned in the turn on the west end of Schwartz Road. It remained stationary during all tests. Figure 3-1 shows Trailer #2 parked on the west end of Schwartz road during a test. The AirPair 100 unit is mounted on the pole on the left.

Trailer #1 was repositioned to the east side on Schwartz Road as needed to set the desired link testing distance. Figure 3-2 shows Trailer #1 parked off to the side of Schwartz road during one of the tests. The AirPair 100 unit is mounted on the pole to the right and is focused up Schwartz Road to the west.

Figure 3-1 AirPair 100 On Trailer #2 At West End Of Schwartz Rd.

Figure 3-2 AirPair 100 On Trailer #1 At East End Of Schwartz Rd.

3.6.2 Shuttle Landing Facility (SLF) Test Setup

The trailers with the E-R Wi-Fi equipment were placed as far apart as possible at the SLF. Fortuitously, the runway length plus the paved overruns at each end provide a separation distance of 3.2 miles. As the goal was to test performance at a distance of 3 miles or slightly greater over a controlled surface roughness path, the SLF provided an extremely uniform path at exactly the distance desired, over which to conduct testing.

As expected, for the low fixed height of the antennas at only 2 meters above the concrete, the predicted Fresnel Zone degradation of received signal strengths kept the link from performing well while testing at this distance, despite the use of otherwise adequate antenna gains and transmitter powers as predicted for closing the link. This result confirmed that the expected Fresnel Zone degradation exists at distances of 3.2 miles with E-R Wi-Fi antennas mounted at a 2 meter height, and that for operation at this distance, positioning of antennas at greater Height Above Average Terrain (HAAT) is recommended.

3.7 TEST EQUIPMENT AND SOFTWARE

Key E-R Wi-Fi test hardware included the following:

- DragonWave AirPair 100 Modem, Radio, and Antenna
- Laptop computer for interfacing with the Modem

In addition to the above hardware, one software utility was instrumental in testing data throughput:

- WSTTCP (a Windows socket port of the BSD TTCP, i.e., of the Unix BSD Test TCP, software) to transfer data through a TCP data flow between the two laptops WSTTCP Ref.s: http://www.pcausa.com/Utilities/pcattcp.htm and http://www.winsite.com/bin/Info?5896

3.7.1 DragonWave AirPair 100

The DragonWave AirPair 100 was the primary component under test. The factory units were modified as shown in Figure 3-3 to include a riflescope attachment. The modem shown in Figure 3-4 was attached to the same mounting pole. Specifications for the units are summarized in Table 3-2. A pair of AirPair 100 units were purchased around 3/1/05 under the ECT task order (#00087).

Included for the purchase price were the following system components:

- (2) AirPair100 Outdoor Radios
- (2) AirPair100 Modems with 30 cm Antennas
- (2) Power supplies
- (2) Power + Ethernet Cables
- (2) IF Cables (Modem – Radio)
- (2) Mounting Brackets
- Link Budget utility
- Factory training for project personnel

Figure 3-3   Modified AirPair 100 Radio with 30cm Antenna & Added Riflescope

Figure 3-4 AirPair 100 Outdoor Modem

Factory training was included with the purchase of the AirPair 100. Training took place in Ottawa, Ontario on 7/14/05. ASRC employees Dr. Gary Bastin, Bill Harris, and José Marín attended.

The units were shipped with a cable assembly that connects to the modem and which breaks out three different interfaces: power, data, and management. The power cable (black) has a 5-pin DIN connector that mates to the power supply provided with the unit. The data cable (blue) provides the 100BaseT Ethernet interface. The management cable (gray) provides the 10BaseT Ethernet interface. The cable assembly is shown separately in Figure 3-5 and connected to the Dragonwave unit in Figure 3-6.

During normal operation, the units can be managed from a Simple Network Management Protocol (SNMP) Manager. Connectivity to the SNMP Agent can be achieved via the data network (in-band) or via the management network (out-of-band). Selection of the network in use for SNMP traffic is user-selectable during system configuration.

Table 3-2   AirPair 100 Specifications

Manufacturer	DragonWave Inc. 600-411 Legget Drive Ottawa, Ontario, Canada K2K 3C9
Model	AirPair 100-UL 24 GHz
Data rate	100 Mbps full duplex performance
Distance	Up to 5 miles
Frequency Range	24.05 – 24.25 GHz
Antenna Diameter	30 cm
Antenna Gain	35.3 dBi
Antenna Beamwidth	2.6°
Modulation	16 QAM
RF Power	+3 dBm
Data Interface	100BaseT
Data Interface Connector	RJ-45
Management Interface (Out-of-band)	10BaseT
Management Interface Connector	RJ-45
Local Management Interface	RS-232
Local Management Interface Connector	DB-9
Voltage (110V adapter supplied)	48 vdc
Environment Operating Temp	-40°F to +122°F
Max Operating wind	70 mph (Antenna)
Warranty	1 year
Serial Numbers AirPair 100 #1	DW320759
Radio #1	410141
AirPair 100 #2	DW320720
Radio #2	410590
Software Loaded	omni_3.4.5.hex

Configuration (and management) of the system can be performed via the Web Interface (available in-band or out-of-band), or via the Command Line Interface (CLI) which is available via telnet (in-band or out-of-band) or via the serial interface (RS-232). Test configuration parameters are listed in Table 3-4.

Figure 3-5 Power + Ethernet Cable

Figure 3-6 Pole Mount with AirPair 100 Modem, Radio, and Antenna

Table 3-3   AirPair 100 Test Configuration Parameters

	East End (#1)	West End (#2)
Software Loaded	omni_3.4.5.hex	omni_3.4.5.hex
Super User name	dw1	dw2
Super User password	dwx	dwx
IP address	128.217.108.180	128.217.108.181
Subnet Mask	255.255.0.0	255.255.0.0
Radio Band	un24	un24
Frequency Bank	Go	Return
Programmed Freq	UNL1 (Tx=24080 Rx=24160)	UNL'1 (Tx=24160 Rx=24080)
Antenna Diameter	30 cm	30 cm
Transmit Power	3 dB	3 dB
Antenna's height	+- 75 inches	+- 75 inches
Laptop IP address	128.217.108.175	128.217.108.174
Auto negotiation	Disabled/Enabled	Disabled/Enabled

3.7.2   **Laptop**

A pair of Gateway laptop computers, purchased in an earlier project year and depicted in Figure 3-7, were used to support E-R Wi-Fi testing. The configuration and management functions performed on the AirPair 100 system were accessible via Internet Explorer, telnet, and Command Prompt (i.e., no special software was required to be loaded to perform these tasks in addition to the Windows Operating System) on the laptops. The link budget software tool provided by DragonWave is simply a Microsoft Excel spreadsheet, built-in to the normal MS Office tools. To perform data throughput testing, each laptop was also loaded with the WSTTCP utility, developed as part of the Open Source software initiative among the Internet community. Both of these utilities are discussed in more detail in a later section.

3.7.3   **DragonWave AirPair 100 Controller Software**

The AirPair 100 can be configured and controlled using the SNMP Manager, Command Line Interface (CLI), or Web Interface.

The SNMP Manager option was not used since that utility is not readily available in the Range lab. Testing via this option will need to be performed among later tests, as part of the final networking compatibility testing, prior to "going operational" with this hardware.

Figure 3-7   Laptop connected to the AirPair 100 Modem

The CLI is accessible three different ways: terminal emulator (RS-232), telnet (Ethernet), or via a custom utility running on a Personal Digital Assistant (PDA) (RS-232). The PDA option was not used since a PDA with the required operating system version was not readily available. The terminal emulator included with the Laptop's operating system (HyperTerminal) was used to perform initial configuration selection of parameters for the AirPair 100 Modems. The telnet utility included with the Laptop's operating system was used to test the interface. This interface is similar to using HyperTerminal.

Using the CLI was difficult since the user is expected to be knowledgeable with the complete command set. Also, this mode requires lots of keyboard use which can be inconvenient when operating the system in the field during field testing.

Figures 3-8 and 3-9 provide screen capture images of two "get" commands (get status of led's and get hardware revision), and also gives a good feel of the operator actions required for using this mode:

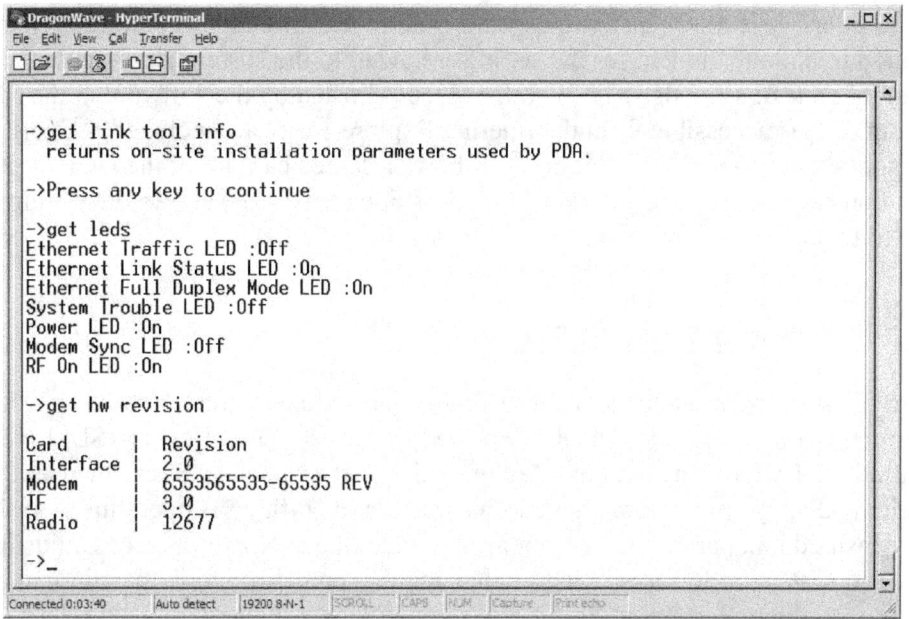

Figure 3-8 CLI screen capture (Laptop)

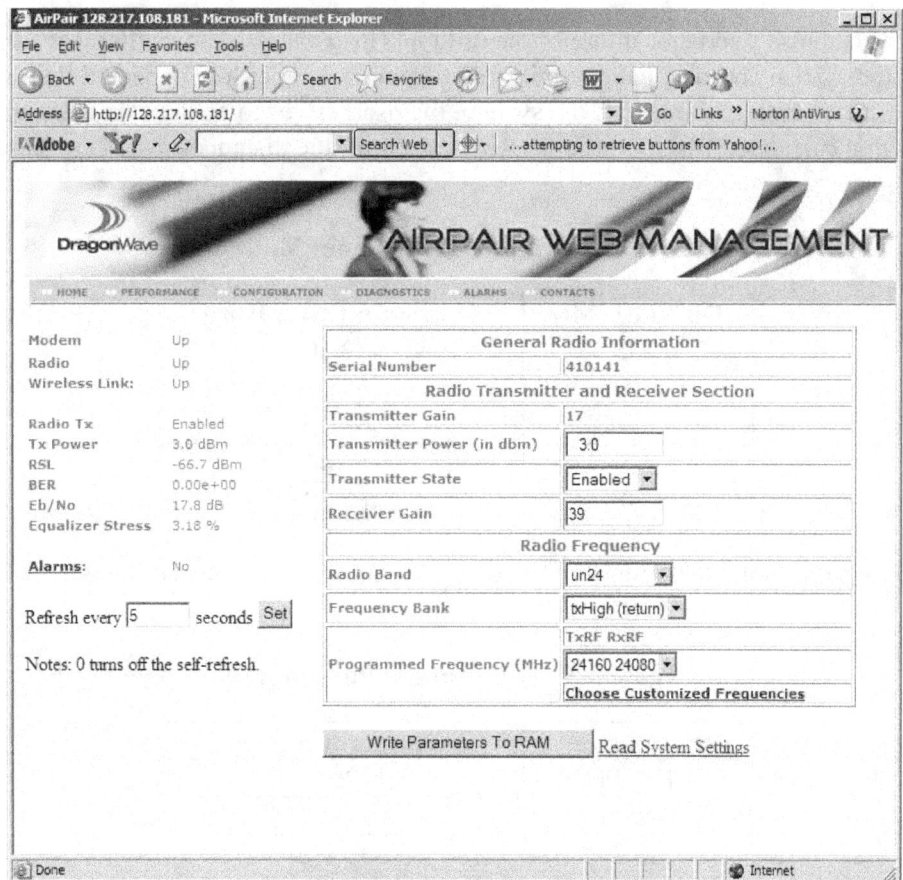

Figure 3-9 Web Interface – General Radio Information Page

The Web Interface is much easier to use and is more user friendly. In addition, it can be configured to poll the modem for the settings relevant to the system alignment at a configurable rate of seconds. The Web Interface is hosted by the software in the Modem, thereby making it accessible from the Internet Explorer software included natively with the Laptop's operating system. Figure 3-9 shows a screen capture of the General Radio Information page of the Web Interface. The left side of the page shows the settings that are polled to simplify operation of the system while in the field.

3.7.4 DragonWave Link Tool

The AirPair 100 system included a software utility to calculate the link budget. This tool enables the user to easily calculate the expected Received Signal Level (RSL) for a given configuration. By knowing the expected operating performance values, the user thus knows immediately upon establishing a link whether everything is operating as it should. When measured link parameters are not as predicted, then something needs adjusting, to achieve full system performance. Through using this prediction tool, the elimination of improperly configured or improperly performing links is achieved, reducing the need to return to the field to re-adjust a link.

Figure 3-10 shows a screen capture of the utility. This example shows that when the two ends of the system are installed one mile apart from each other, the expected RSL is -52.10 dBm. During alignment of the system, the user strives to achieve an RSL within 3 dBm of this calculated RSL. Optimum RSL values for the distances tested are shown in Table 3-4.

Table 3-4   Optimum RSL Values Vs Distances

Distance (Mi)	Optimum RSL (dBm)
1.0	-52.10
1.5	-55.75
2.0	-58.39
2.5	-60.46
3.0	-62.18

Other optimum parameters monitored during setup are shown in Table 3-5.

Table 3-5   Optimum Parameters

Parameter	Optimum Range
Eb/No	>20.0
Stress	3-5%

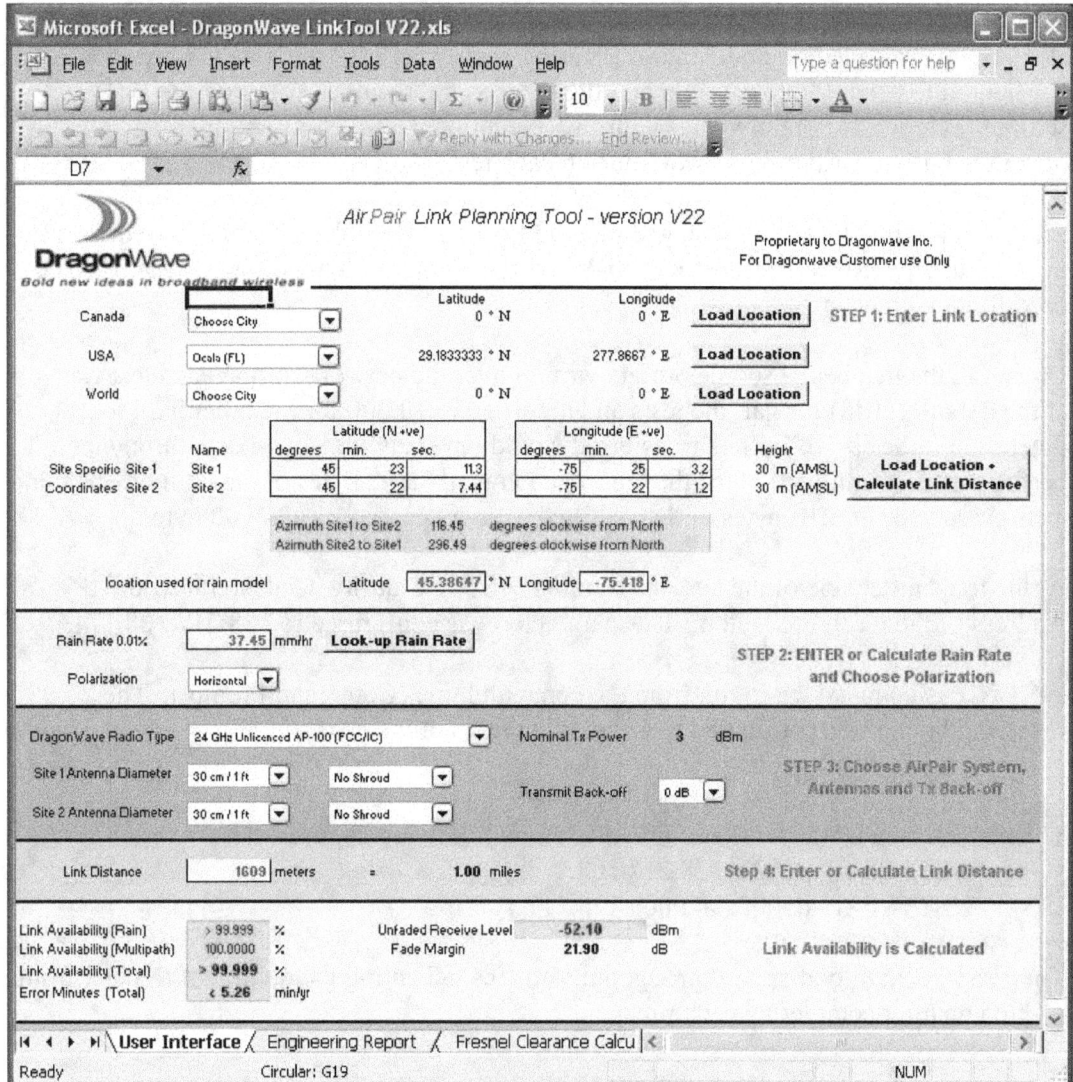

Figure 3-10   LinkTool utility

3.7.5 Throughput Testing

Throughput testing was performed with the help of the "WSTTCP" utility.

> WSTTCP Ref.s:
> http://www.pcausa.com/Utilities/pcattcp.htm and
> http://www.winsite.com/bin/Info?5896

WSTTCP is a WINSOCK (i.e., a Windows socket) port of the BSD TTCP, i.e., of the Unix BSD Test TCP, software. It is used to transfer data through a TCP data flow between two computers.

WSTTCP was loaded on both laptops. During each WSTTCP run, one laptop was configured to act as the transmitter and the other as the receiver. Most of the WSTTCP settings were left at default values with two exceptions: socket port number and number of source buffers written to network during each test.

The socket port number was changed to 5002 (from 5001). This change was required as the two laptops' Norton Symantec Anti-Virus software would not allow port 5001 to be used for data transfer purposes.

Likewise, the number of source buffers written over the network during each test was left at the default (2048) for half the tests and modified to 10,000 for the other half, to determine if the size of buffers transferred would have any impact on data throughput performance. During each test, the buffers contained 8,192 bytes of data. In effect, some tests passed 16,777,216 bytes of data while the others passed 81,920,000 bytes.

At the transmitter side of the test, the running WSTTPC utility needs to know the IP Address of the receiving laptop. This parameter was configured to 128.217.108.174.

WSTTCP is commanded to run from the command line (Command Prompt). The receiving laptop was configured with the following command:
    WSTTCP –r –p5002

The transmitting laptop was configured with the following commands:
    WSTTCP –t –p5002 128.217.108.174
    WSTTCP –t –p5002 –n10000 128.217.108.174

The WSTTCP utility displays throughput statistics indicating the quality of the link from end to end after completing each run.

3.8   TEST RESULTS

3.8.1   General

E-R Wi-Fi testing started at the Engineering Development Lab (EDL) where the AirPair 100 systems were installed on the trailers and powered-up for the first time. The primary test objective at this location was setup, initialization, checkout, and familiarization, prior to testing at a remote site where there would be few resources to debug problems.

The second stage of testing took place at Schwartz Road. At this location, the system was tested two different times. This first time it was tested at distances of 1.0, 1.5, 2.0, and 2.5 miles apart. After completing these tests, the test engineers attended a vendor-provided training course. Testing resumed shortly after returning from the training. This time, the system was tested at distances of 1.0, 2.0, and 2.5 miles apart. The primary test objective at this Schwartz Road location was making data throughput measurements over the wireless network provided by the AirPair 100 system. The 24 GHz absorbing rough terrain vegetation alongside Schwartz Road largely avoided the deleterious effects of Fresnel Zone degradation during these tests.

As not all fielded sites will be operating over absorptive materials, the effects of Fresnel Zone degradation due to destructive interference between direct and reflected signals off the ground was desired to be verified. Published rules-of-thumb nomograms indicated that such degradation would likely occur around 3.0 miles, or slightly greater, provided a uniformly consistent and largely non-disappative surface was beneath the link, for the height above ground of 2 meters for the mounted microwave dish antennas. Hence, the third stage of the testing took place at the Shuttle Landing Facility (SLF), as this would enable determining what the worst case degradation for a very uniform surface beneath the propagating path would be. At this location the primary objective was to push the distance limit of the system to slightly more than 3.0 miles, and see if link degradation above what the previously mentioned link tool predicted, would occur. It was expected to experience difficulties obtaining a link with a signal quality capable of providing a reliable path for a successful data throughput test, despite the use of adequate antenna gain and transmitter power and receiver sensitivity to otherwise close the link. The maximum capable distance at the SLF is 3.2 miles of separation. It was decided to push the equipment to the full 3.2 miles.

After reviewing the data collected through the first 3 stages, DragonWave field support was contacted and helped mitigate an obvious shortcoming that we noticed in the throughput performance for our system (e.g., data throughput was never achieving higher than 8.6 Mbps). It was noted that our AirPair 100 system was improperly configured to keep the network interface from automatically negotiating the network's speed (improperly setting operation at 10Mbps vs. 100Mbps). This setting kept the test laptop from operating at its full speed of 100Mbps. A fourth set of tests was added to the testing to enable testing throughput while simultaneously allowing the system to negotiate the network's speed automatically, and thereby achieve the highest data rate throughput performance possible with the equipment. Table 3-6 summarizes the types of tests.

Table 3-6   **Summary of Test Locations & Objectives**

Location	Test Objectives
EDL	Setup, initialization, checkout, and familiarization
Schwartz Road	Familiarization and data throughput
SLF	Push the distance limitation
Schwartz Road	Test data throughout with Auto Negotiate Enabled

3.8.2   Results Summary

E-R Wi-Fi testing was conducted from May 2005 through September 2005. The first phase (setup, initialization, checkout, and familiarization) was performed at the EDL parking lot at a distance shorter than what the system was designed to operate (approximately 60 meters). During this phase DragonWave's tech support was very helpful and was readily available to address the initial issues and noted concerns. During this phase, no data throughput testing was performed, as the topic of interest was simply hardware familiarization and proper usage checkout.

The second phase was conducted at Schwarz Road from July 2005 to mid-August 2005. Successful links were established at distances of 1.0, 1.5, 2.0, and 2.5 miles. The mounting hardware used for the antennas proved to be overly difficult to use during alignment. Once aligned, this hardware secures the antenna in place as it was designed to do. Unfortunately, this hardware is not designed for quickly aligning test setups regularly, but rather for rarely aligning fixed installations. We added a 12x power riflescope mount to expedite quick repeated alignments on a recurring basis, as testing configurations were modified. Testing at Schwartz Road provided a fairly easy way to align the antennas since the road is basically a straight line.

The data throughput values recorded during the second phase showed that the system had a 10 Mbps limitation. This was attributed to the fact that the AirPair 100 units were configuring themselves improperly with the network speed auto-negotiation disabled. These units default to 100 Mbps when auto-negotiation is disabled, but the laptop defaults to 10 Mbps, thereby constricting data throughput to the lower of the two, 10 Mbps, with auto-negotiation disabled. This default configuration worked against achieving the best possible overall system performance.

The third phase was conducted at the SLF in late August 2005. Pushing the limits of the link to this distance over a very uniform path surface confirmed that the expected Fresnel Zone degradation exists at a distance of 3.2 miles with the E-R Wi-Fi antennas mounted at a low 2-meter height. The best received signal level (RSL) achieved during this test had a link performance unacceptable to perform a data throughput measurement test. For this reason, no throughput data was collected during this phase, despite the link tool theoretically indicating that the link should be capable of closing (passing data.)

For phase four the unit's were configured with the network speed auto negotiation enabled. The system was returned to Schwarz Road to verify that this setting was limiting the data throughput. Initially the units were placed 2.5 miles apart. The system performed as expected (data throughput averaging around 70 Mbps). The units were then tested at distances of 2.0, 1.5, and 1.0 miles apart, and data rates much closer to the theoretical maximum of 100 Mbps were achieved and measured.

Figure 3-11 shows all the data collected during the second and fourth phases of field testing. It shows throughput in Mbps over the distance (in miles) separating the two ends of the AirPair 100 system.

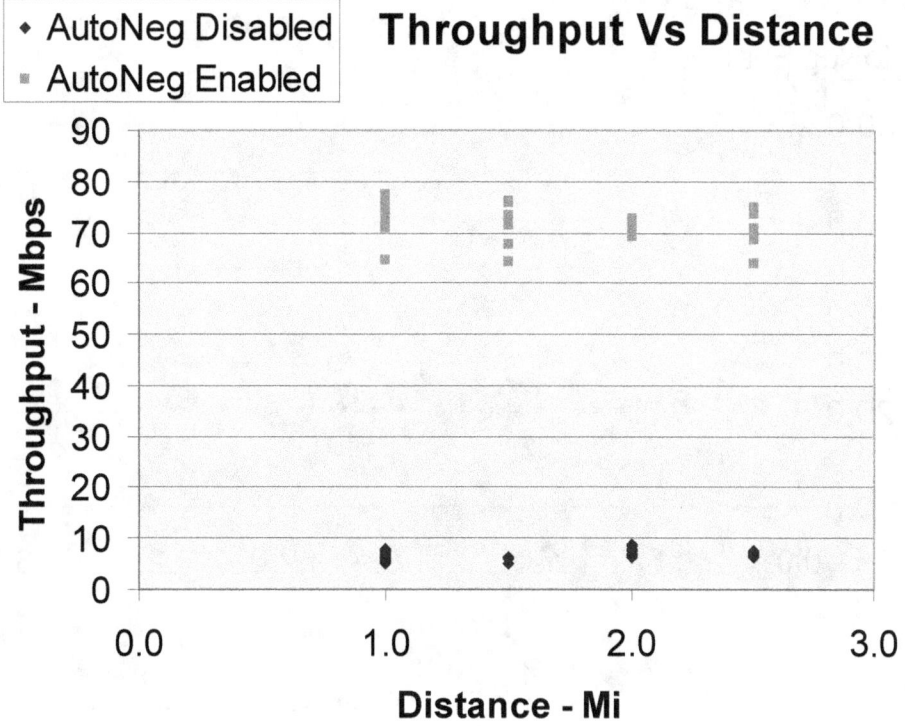

Figure 3-11 Throughput Versus Distance

Figure 3-12 shows the calculated average data throughput for the same distances.

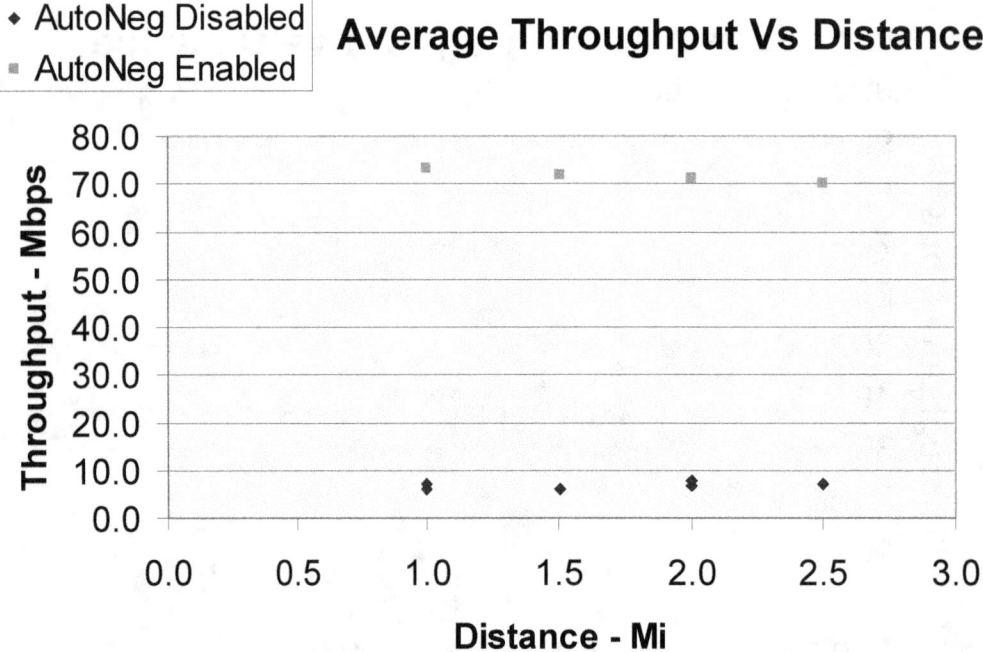

Figure 3-12   Average Throughput versus Distance

3.9   E-R WI-FI SECURITY CONCERNS

Security for typical Wi-Fi links is usually achieved using WEP, or newer security protocols. As the emphasis this year was on simply achieving communication over longer range distances than possible with Wi-Fi, at locations for which physical security through zone of control would be adequate, additional protocols for securing E-R Wi-Fi links were not investigated. In short, the usual protocols would provide as much security as for shorter wide area networks utilizing wireless access points, and no additional security was deemed necessary.

For fielded links, operating where zone of control limiting undesired intruders is not possible, more secure protocols will need to be employed. This is recommended for a future set of tests, during the initial testing in an operational configuration during field testing. This testing is planned for the coming year.

3.10   E-R WI-FI SUMMARY AND RECOMMENDATIONS

The precept of ECT during Phases 2 and 3 was to investigate newly-introduced Wi-Fi, UWB, and FSO products while they were still in their infancy, and thereby influence the development of these developing products as early as possible, before we needed operational systems on the Range at KSC. For both Wi-Fi and Extended-Range Wi-Fi, we are now much closer to fielding practical systems at KSC, and we have clearly gained an in-depth understanding of the operational limits of extended range Wi-Fi technology within the unique KSC environment. As in any activity, the state of the art continues to improve. And we have found that there are critical unique aspects of the KSC environment that force a shift from the priorities that commonly exist for Wi-Fi systems intended for urban areas.

For example, in urban environments, or in densely occupied office environments, where many users can be located within a common area within range of a single wireless access point, the need for strong security protocols is clearly mandated.

However, within the Range environment of KSC, over long distance areas for which zone of control is clearly adequate to protect operational links, less strong security protocols are entirely feasible for the next few years for initial operational testing. Clearly, though, stronger protection against intrusion and surreptitious use of E-R Wi-Fi links will be required for networking remote sites at KSC with the existing wired infrastructure, and this protection must be developed and tested using stronger protection protocols than the existing WEP and similar techniques. These topics must be addressed in the next set of activities, before transitioning to fielded operational E-R Wi-Fi systems.

We are clearly much closer to our goal of fielding high data-rate rates over longer distances at KSC remote sites with Wireless Ethernet (Wi-Fi) systems.

4.0

ULTRA WIDE BAND

4.1 INTRODUCTION

Emerging Ultra Wideband (UWB) systems hold the promise of delivering wireless data at high speeds, exceeding hundreds of megabits per second over typical distances of 10 meters or less.

Last year's ECT-related UWB activities investigated the timing accuracies required for achieving low Bit Error Rates despite the presence of timing errors which can otherwise adversely affect the positional accuracies and Bit Error Rates of UWB systems. The prior year's ECT-related UWB activity provided an investigation of the methods of detecting non-cooperative UWB signals over distances that are essentially as far as the actual communication distances usable by UWB receivers utilizing all detailed waveform parameters, and resulted in a patent application.

The emphasis in this year's activities involves optimizing UWB waveforms for achieving overall system power consumption efficiency, and additionally monitoring industry progress towards achieving the necessary foundation infrastructure needed to make UWB system components practical.

4.2 BACKGROUND

In the near future, wireless broadband communications systems will require data rates exceeding hundreds of mega bits per second (Mbps). To address these approaching demands, emerging Ultra Wideband (UWB) systems offer an ideal physical (PHY) layer solution to address wireless personal area networking (WPAN) needs over short ranges.

As UWB modulation becomes better understood and data rates increase to near their maximum potentials, the emphasis will likely switch to maximizing efficiency of UWB systems through optimizing UWB waveforms to accommodate transmitting at lower transmitter power levels.

The ECT Phase 4 research this year explored the waveform details required to support the operation of UWB OFDM systems in such a future communication landscape. Additionally, the activities this year tracked the progress among competing camps proposing a unifying UWB alternative standard for implementing an emerging communication technology. Unfortunately, the UWB task group (802.15.3a) still remains deadlocked, where it has remained since 2003, with two incompatible approaches. This is/was the standard that was supposed to enable cable TV without the cable and permit interfacing between flat panel HD TVs and HD playback recorders, as well as between cable TV inputs into homes and remote wall-mounted flat-panel HD TVs, all without the wires and cables required to date.

4.3   UWB WAVEFORM INVESTIGATION UPDATE

Instead of using the traditional Pulse Position Modulation (PPM), Pulse Amplitude Modulation (PAM), Binary Phase Shift Keying (BPSK), and On/Off Keying (OOK) modulations investigated in the first two years of UWB research, and the alternate approach investigated last year of OFDM (Orthogonal Frequency Division Multiplexing), the emphasis this year is instead to focus on the physical layer (PHY) and investigate fundamental techniques for improving the power efficiency of future UWB systems through investigating choice of waveform optimization techniques.

Why does any particular choice of monocycle (UWB) waveform reduce power consumption? It is really not that less data is lost, or that the computational power is reduced through having a more optimal monocycle waveform. Instead, it is because:

1. For any given modulation selected, the signal to noise ratio (SNR) is what most determines the Bit Error Rate (BER), provided that timing errors are otherwise minimized (the effects of which were analyzed last year), interference is otherwise minimized, and jamming signals are not present.

2. Some types of modulation are more robust than others, that is, they provide better BER performance at low SNR, when operated in thermal noise-limited environment.

3. A monocycle is very close to being the optimal Gaussian signal that can provide the most robust signaling scheme in an AWGN (Average White Gaussian Noise) environment, thereby providing the very best BER at low SNR, as noted in the activities of two years prior.

4. Because of the ability of a monocycle to operate at the lowest SNR while providing the best communication BER, a communication link designed around UWB monocycles is theoretically capable of being the most efficient for a given transmitted power in terms of BER performance. Equivalently, for achieving a given BER, it becomes possible to use a monocycle waveform to permit using the least amount of transmitted power of any other modulation type. (Not just known modulation formats, but of all possible modulation formats!)

5. Relative to where power is most consumed, transmitters generally use the most power; hence, for many modulation types, the amount of DSP chip power used for demodulation in a receiver is roughly constant, ceteris paribus. It is the transmitter power that matters most in terms of power consumption efficiency.

6. Because of this optimal signaling characteristic, a UWB waveform based on a monocycle uses the least amount of transmitted power, and hence provides the most power-efficient communication link that is theoretically possible.

7. This advantage is an even bigger deal when networks made up of large numbers of sensors are used. Such a network is most able to benefit from UWB's optimal signaling advantage, as there is typically only one receiver, and a very large number of transmitters. UWB is inherently more of an advantage over other modulation types for optimizing power consumption in such a wireless sensor grid. It would also become possible to have battery-powered UWB sensors that would permit total freedom from the ac power mains in many applications when operation over only a few weeks or less is desired from the UWB wireless sensor grid transmitters. The receiver, of course, could be centrally located where ac power was available.

Based on these observations, research activities with Ms. Tammara Massey, NASA Harriett Jenkins Fellowship student and UCLA graduate student, were commenced this year. Ms. Massey is continuing her research into the next fiscal year, and this research will be continued, in a joint effort, among ECT personnel and Ms. Massey. It is anticipated that an MSEE thesis will ultimately result from this jointly-conducted research work. Ms. Massey, however, is not utilizing any budget from ECT, but is instead funded through the NASA Harriett Jenkins Fellowship program.

4.4   STANDARDS UPDATES

A major UWB news announcement was released on May 9, 2005, and continues to build; UWB and Bluetooth are effectively merging into a new standard. The newly-merged standard is to be known as 802.15.4a.

This merger of standards will provide an upgrade path for Bluetooth, a short range WPAN comm technology that is currently locked to a maximum data rate of roughly 3 Mb/s, and will enable it to achieve data rates of 480 Mb/s and higher through integrating a second PHY layer implemented with UWB into the next version of Bluetooth.

It is not clear if this is a viable approach to achieving higher performance, or if it is merely the last faint gasp of two technologies that have seen their fortunes decline somewhat over the past 2 years due to the failure of the market to settle on a particular UWB standard and due to the rapid growth of Wi-Fi and related similar wireless technologies (Wi-Max, etc.) at the expense of Bluetooth -- Bluetooth market share has never done well in North America, although it has done better in Europe. In North America, its largest use has been capturing the hands-free earpiece comm link for GSM cellphones, building on technology first introduced in Europe for their GSM phones. Meanwhile, UWB remains a "future technology" due to its lack of a unifying compatible standard, among either IEEE committees and FCC recognized regulations..

4.5   UWB ANTENNA UPDATES

Significant advances were made this year among UWB industry participants into developing wider bandwidth UWB antennas. Perhaps most significant, Fractus, of Barcleona, Spain, announced in early September 2005 the release of a UWB antenna design dubbed the UWB Media+ Chip Antenna. This antenna operates over the 3.1-5.0 GHz sub-banded spectrum specified by the WiMedia Alliance, which includes members Intel, Texas Instruments, Microsoft, Philips Electronics, Nokia, and Staccato Communications among others. The WiMedia Alliance is one of the two consortia competing for dominance among the UWB task group (802.15.3a). Staccato claims to have tested the fractal-based antenna and, based on successful tested performance, began offering it as an optional antenna for use with its all-CMOS UWB chip. Fractus' UWB antenna measures 10.0 mm x 10.0 mm x 0.8mm, and is claimed to provide an omni-directional radiation pattern at high efficiency. [18]

4.6   SUMMARY OF RESULTS

The primary UWB-related activity this year was simply monitoring the rather slow progress in the vendor community and among standards committees for UWB technology, while additionally conducting joint research through the NASA Harriett Jenkins Fellowship program with NASA Fellowship student Ms. Tammara Massey. Due to the uncertainty of which of the two competing UWB approaches may ultimately win the majority of market share, the release of new designs from vendors was noticed to have slowed noticeably this year, since the initial flurry of activity two and three years ago.

Nonetheless, UWB remains a most intriguing technology, with communication capabilities that largely have been under-appreciated among the general community of First Mile/Last Mile communication link planners.

4.7   RECOMMENDATIONS FOR FUTURE RESEARCH

Recommendations for continuing UWB research in the new fiscal year include investigating, selecting, procuring, and testing new UWB Evaluation Kits released in 2005, one of which may help crystalize the market into settling on a de-facto industry standard.

Likewise, the research into optimal power-saving UWB waveforms will be continued at no or low cost to the project through utilizing the resources of the NASA Harriett Jenkins Fellowship student program, through working with UCLA graduate student Ms. Tammara Massey.

This way, UWB technology will be monitored and best tailored for future use on the Range.

---

[18] "Fractal antenna pioneer targets UWB", John Walko, EE Times, 09/13/05, http://www.eetimes.com/news/latest/showArticle.jhtml?articleID=170702765

5.0   ECT SUMMARY RECOMMENDATIONS FOR CONTINUED RESEARCH

The following major task activity areas are recommended for continued research in the next fiscal year for emerging communication technology development:

1. Networked FSO
2. FSO pre-operational field testing during inclement weather
3. Extended-Range Wi-Fi pre-operational field testing during inclement weather.
4. UWB Evaluation Kit (EVK) procurement and testing.
5. Monitor emerging communication technology development.

Within these five major task activity areas, the following detailed tasks are recommended:

1. Networked FSO:
- Conduct an industry survey of FSO system hardware suitable for implementing a small networked optical FSO system, and procure a fixed, non-tracking, wide-beam optical FSO system set of components with redundant optical beams to investigate the limits of networking multiple FSO systems.

2. FSO Pre-operation Field Testing:
- Update previously-generated test procedures, while adapting and expanding these procedures to account for the multiple FSO OTUs.
- Test the FSO system exemplars for Bit Error Rate, and throughput rates versus weather-induced degradation conditions (e.g., fog, rain, etc.) when operating in a networked configuration.
- Test the performance limits of FSO hardware within the unique environment of KSC, with data path links over both water and over land, comparing the applicability of this technology to KSC's needs versus the single FSO link tested previously, including during inclement weather.

3. Extended-Range Wi-Fi Pre-Operational Field Testing
- Update previously-generated test procedures, while adapting and expanding these procedures to account for the longer distance Wi-Fi links.
- Test the E-R exemplars for Bit Error Rate, and throughput rates versus weather-induced degradation conditions (e.g., fog, rain, etc.).
- Test the performance limits of E-R Wi-Fi hardware within the unique environment of KSC, with data path links situated over both water and over land, comparing the applicability of this technology to KSC's needs versus the shorter-range Wi-Fi links tested previously, including during inclement weather.

4. UWB Evaluation Kit Procurement and Testing:
   - Conduct an industry survey of UWB EVK hardware suitable for implementing a short range (<10 m) communication system, and procure an EVK set of components with a data rate of several hundred Mb/s. Update previously-generated test procedures, while adapting and expanding these procedures for testing the UWB evaluation kit for position aware functionality.

5. Monitor Emerging Communication Technology Development
   - Review current UWB and FSO products and theoretical developments through attending two major communication conferences.

These task order activities are needed to achieve the 24/7, always-on, highly-mobile vision of an interconnected communication for use on the Range employing First Mile / Last Mile extensions to the existing Range communication infrastructure.

APPENDIX A: Acronyms & Definitions

APD	Avalanche Photodetector Diodes
ARTWG	Advanced Range Technologies Working Group
ASTWG	Advanced Spaceport Technologies Working Group
AWGN	Average White Gaussian Noise
BER	Bit Error Rate
BPSK	Binary Phase Shift Keying
BSD	Berkeley Software Distribution from University of California at Berkeley; also known as BSD Unix
BW	Bandwidth
CDR	Clock and Data Recovery
CEV	Crew Exploration Vehicle
COTS	Commercial off the shelf
dB	decibel; formally, it is 10 times the logarithm to base 10 of the ratio of two powers
dBe	decibel electrical
dBo	decibel optical
DSP	Digital Signal Processing
DW1	DragonWave unit #1
DW2	DragonWave unit #2
E_b	Energy per bit
ECT	Emerging Communication Technologies
EDL	Engineering Development Laboratories (building)
E-R WiFi	Extended-Range WiFi
FCC	Federal Communications Commission
FEC	Forward Error Correction
FIRST	Future Interagency Range & Spaceport Technologies

FSO	Free Space Optics
FY	Fiscal Year
GSE	Ground Support Equipment
GSM	Global System for Mobile Communication; originally a lengthy French acronym shortened to **GSM** from "*Groupe de travail Spéciale pour les services Mobiles*", but which is now generally considered Anglicised
HAAT	Height above average terrain
Hz	Hertz (cycle per second)
IFFT	Inverse Fast Fourier Transform
IM/DD	Intensity Modulated / Direct Detection
IP	Internet Protocol; used with Transmission Control Protocol (TCP), TCP/IP forms the rule set that enables computers to communicate via the Internet
MBOA	Multi-Band OFDM (Orthogonal Frequency Division Multiplexed) Alliance; one of the two competing UWB industry partnerships
Mbps	Megabits per second; 10^6 bits per second
Mux	Multiplexer
NASA	National Aeronautics and Space Administration
N_{eff}	Effective noise
OE	Optical to Electrical
OFDM	Orthogonal Frequency Division Multiplexing
OOK	ON/OFF Keying
OTU	Optical Transfer Unit
PAM	Pulse Amplitude Modulation
PAT	Performance Analysis Tool
PDF	Probability Density Function
PHY	Physical Layer
PIN	Positive-Intrinsic-Negative (diodes)
PPM	Pulse Position Modulation

PRBS	Pseudo-Random Bit Sequence
R&D	Research & Development
RF	Radio Frequency
RISM	Range Information Systems Management
Rx	Receiver
SBRDS	Space Based Range Distributed Subsystem
SLF	Shuttle Landing Facility
SNR	Signal to Noise Ratio
<SNR>	Time average value of Signal to Noise Ratio
SPIE	Society of Photo-Optical Instrumentation Engineers, also known as "SPIE – The International Society for Optical Engineering"
TCP	Transmission Control Protocol; used with Internet Protocol (IP), TCP/IP forms the rule set that enables computers to communicate via the Internet
Tx	Transmitter
UI	Unit Interval
UK	United Kingdom
Unk	Unknown
UWB	Ultra Wide Band
WiFi	Wireless Fidelty
WSTTCP	**W**indows **S**ocket port of **T**est **T**ransmission **C**ontrol **P**rotocol software ported originally from BSD operating system TTCP utility by Sungjin Chun <sjchun@janus.sst.co.kr> to Windows Sockets as WSTTCP in 1996
WPAN	Wireless Personal Area Network
Wx	Weather
<SNR>	Time average value of the Signal to Noise Ratio

APPENDIX B: Optical BER Equations For OOK

The difference in dBe (electrical decibels) versus dBo (optical decibels) primarily comes from the optical to electrical conversions occurring in photodiodes. A similar difference also occurs when deriving theoretical BER equations for RF vs. Electro-Optical OOK systems in terms of signal-to-noise ratios.

To see this difference, let the output current from the receiver photodiode be (neglecting atmospheric effects):

$$i = i_s + i_n \qquad (11)$$

where i_s = signal current and i_n = noise current. Hence, $<i> = i_s$ and $\sigma_i^2 = \sigma_n^2 = <i_n^2>$. If we assume the noise to be zero-mean Gaussian, then:

$$p_n(i) = \frac{1}{\sigma_n \sqrt{2\pi}} \exp\left(-\frac{i^2}{2\sigma_n^2}\right) \qquad (12)$$

$$p_{s+n}(i) = \frac{1}{\sigma_n \sqrt{2\pi}} \exp\left(-\frac{(i-i_s)^2}{2\sigma_n^2}\right). \qquad (13)$$

The first expression is the pdf on noise alone and the second is the pdf for signal and noise. Hence, the false alarm probability and detection probability are given by:

$$P_{fa} = \int_T^\infty p_n(i)di = \frac{1}{2}\mathrm{erfc}\left(\frac{T}{\sigma_n \sqrt{2}}\right) \qquad (14)$$

$$P_{det} = \int_T^\infty p_{s+n}(i)di = \frac{1}{2}\mathrm{erfc}\left(\frac{T-i_s}{\sigma_n \sqrt{2}}\right) \qquad (15)$$

where T is the threshold. The error probability (BER) for a digital system is defined by:

$$BER = \frac{1}{2}P_{fa} + \frac{1}{2}P_{fade} \qquad (16)$$

where $P_{fade} = 1 - P_{det}$. For OOK, set a threshold halfway between "ON" and "OFF", i.e., at $T = 0.5 i_s$. In this case, we find:

$$P_{fa} = P_{fade} = \frac{1}{2}\mathrm{erfc}\left(\frac{i_s}{2\sqrt{2}\sigma_n}\right) \qquad (17)$$

$$BER = \frac{1}{2} erfc\left(\frac{i_s}{2\sqrt{2}\sigma_n}\right) = \frac{1}{2} erfc\left(\frac{1}{2}\sqrt{\frac{z}{2}}\right), \quad z = SNR = \frac{i_s^2}{\sigma_n^2} \quad (18)$$

The effect of the optical power to electrical current conversion using a photodiode is to add approximately 6 dB to the theoretical OOK <SNR> required for electro-optical versus RF/electrical communication systems, to achieve the same BER performance.

APPENDIX C: Extended Range Wi-Fi Test Results

The following data were collected during the second stage of testing. The testing was performed at Schwartz Road at distances ranging from 1.0 to 2.5 miles. During all of these tests, the system's network interface auto-negotiation was disabled. DW1 represents the East AirPair 100 unit that was connected to the laptop transmitting the test data. DW2 represents the West unit receiving the test data.

Run	Date	Dist Mi	Bytes	Auto Neg	DW1 Tx dB	DW1 RSL dBm	DW1 Eb/No dB	DW1 Thrpt Mbps	DW2 Tx dB	DW2 RSL dBm	DW2 Eb/No dB	DW2 Thrpt Mbps
1	7/5/05	1.0	81920000	Disabled	3	-51.5	n/a	6.1	3	-47.7	n/a	6.1
2	7/5/05	1.0	81920000	Disabled	3	-51.5	n/a	6.5	3	-47.7	n/a	6.5
3	7/5/05	1.0	81920000	Disabled	3	-51.5	n/a	5.5	3	-47.7	n/a	5.5
4	7/5/05	1.0	81920000	Disabled	3	-51.5	n/a	5.5	3	-47.7	n/a	5.6
5	7/5/05	1.0	16777216	Disabled	3	-51.5	n/a	5.0	3	-47.7	n/a	5.0
6	7/5/05	1.0	16777216	Disabled	3	-51.5	n/a	6.3	3	-47.7	n/a	6.3
7	7/5/05	1.0	16777216	Disabled	3	-51.5	n/a	7.3	3	-47.7	n/a	7.3
8	7/5/05	1.5	16777216	Disabled	3	-59.3	n/a	5.2	3	-56.0	n/a	5.2
9	7/5/05	1.5	16777216	Disabled	3	-59.3	n/a	6.1	3	-56.0	n/a	6.1
10	7/5/05	1.5	16777216	Disabled	3	-59.3	n/a	6.3	3	-56.0	n/a	6.3
11	7/5/05	1.5	81920000	Disabled	3	-59.3	n/a	6.1	3	-56.0	n/a	6.1
12	7/5/05	1.5	81920000	Disabled	3	-59.3	n/a	6.4	3	-56.0	n/a	6.4
13	7/5/05	1.5	81920000	Disabled	3	-59.3	n/a	5.8	3	-56.0	n/a	5.8
14	7/6/05	2.5	81920000	Disabled	3	-62.9	n/a	6.8	3	-59.9	n/a	6.7
15	7/6/05	2.5	81920000	Disabled	3	-62.9	n/a	6.6	3	-59.9	n/a	6.6
16	7/6/05	2.5	81920000	Disabled	3	-62.9	n/a	7.2	3	-59.9	n/a	7.2
17	7/6/05	2.5	16777216	Disabled	3	-62.9	n/a	6.5	3	-59.9	n/a	6.4
18	7/6/05	2.5	16777216	Disabled	3	-62.9	n/a	7.4	3	-59.9	n/a	7.4
19	7/6/05	2.5	16777216	Disabled	3	-62.9	n/a	7.9	3	-59.9	n/a	7.7
20	7/6/05	2.0	81920000	Disabled	3	-63.9	n/a	6.8	3	-59.6	n/a	6.8
21	7/6/05	2.0	81920000	Disabled	3	-63.9	n/a	6.6	3	-59.6	n/a	6.6
22	7/6/05	2.0	81920000	Disabled	3	-63.9	n/a	6.5	3	-59.6	n/a	6.5
23	7/6/05	2.0	16777216	Disabled	3	-63.9	n/a	6.3	3	-59.6	n/a	6.2
24	7/6/05	2.0	16777216	Disabled	3	-63.9	n/a	6.4	3	-59.6	n/a	6.4
25	7/6/05	2.0	16777216	Disabled	3	-63.9	n/a	8.2	3	-59.6	n/a	8.0
26	8/9/05	2.0	16777216	Disabled	3	-61.1	n/a	8.5	3	-58.4	n/a	8.4
27	8/9/05	2.0	16777216	Disabled	3	-61.1	n/a	6.8	3	-58.4	n/a	6.8
28	8/9/05	2.0	16777216	Disabled	3	-61.1	n/a	8.0	3	-58.4	n/a	7.8
29	8/9/05	2.0	81920000	Disabled	3	-61.1	n/a	7.4	3	-58.4	n/a	7.4
30	8/9/05	2.0	81920000	Disabled	3	-61.1	n/a	7.4	3	-58.4	n/a	7.4
31	8/9/05	2.0	81920000	Disabled	3	-61.1	n/a	8.7	3	-58.4	n/a	8.7
32	8/10/05	1.0	16777216	Disabled	3	-52.8	n/a	8.0	3	-48.5	n/a	7.9
33	8/10/05	1.0	16777216	Disabled	3	-52.8	n/a	8.0	3	-48.5	n/a	8.0
34	8/10/05	1.0	16777216	Disabled	3	-52.8	n/a	6.4	3	-48.5	n/a	6.4

35	8/10/05	1.0	16777216	Disabled	3	-52.8	n/a	7.5	3	-48.5	n/a	7.4
36	8/10/05	1.0	16777216	Disabled	3	-52.8	n/a	7.6	3	-48.5	n/a	7.6
37	8/10/05	1.0	81920000	Disabled	3	-52.8	n/a	7.4	3	-48.5	n/a	7.4
38	8/10/05	1.0	81920000	Disabled	3	-52.8	n/a	6.9	3	-48.5	n/a	6.9
39	8/10/05	1.0	81920000	Disabled	3	-52.8	n/a	6.7	3	-48.5	n/a	6.7
40	8/10/05	1.0	81920000	Disabled	3	-52.8	n/a	6.4	3	-48.5	n/a	6.4
41	8/10/05	1.0	81920000	Disabled	3	-52.8	n/a	7.3	3	-48.5	n/a	7.3
42	8/10/05	1.0	81920000	Disabled	3	-52.8	n/a	6.7	3	-48.5	n/a	6.6
43	8/10/05	1.0	81920000	Disabled	3	-52.8	n/a	6.8	3	-48.5	n/a	6.8
44	8/10/05	2.5	16777216	Disabled	3	-63.8	n/a	7.5	3	-58.9	n/a	7.4
45	8/10/05	2.5	16777216	Disabled	3	-63.8	n/a	7.6	3	-58.9	n/a	7.6
46	8/10/05	2.5	16777216	Disabled	3	-63.8	n/a	6.8	3	-58.9	n/a	6.8
47	8/10/05	2.5	81920000	Disabled	3	-63.8	n/a	7.3	3	-58.9	n/a	7.3
48	8/10/05	2.5	81920000	Disabled	3	-63.8	n/a	6.8	3	-58.9	n/a	6.8
49	8/10/05	2.5	81920000	Disabled	3	-63.8	n/a	7.3	3	-58.9	n/a	7.3
50	8/10/05	2.5	81920000	Disabled	3	-63.8	n/a	7.3	3	-58.9	n/a	7.3

This next table shows the data collected during the fourth stage of testing. The testing was performed at Schwartz Road. The distances range between 1.0 and 2.5 miles of separation. During all of these tests the system's network interface auto-negotiation was enabled. DW1 represents the AirPair 100 unit that was connected to the laptop transmitting the test data. DW2 represents the unit receiving the test data.

ECT – Phase 4

Run	Date	Dist Mi	Bytes	Auto Neg	DW1				DW2			
					Tx dB	RSL dBm	Eb/No dB	Thrp Mbps	Tx dB	RSL dBm	Eb/No dB	Thrpt Mbps
95	9/13/05	2.5	16777216	Enabled	3	-62.2	17.1	73.6	3	-56.9	22.3	72.8
96	9/13/05	2.5	16777216	Enabled	3	-62.2	17.1	74.9	3	-56.9	22.3	74.9
97	9/13/05	2.5	16777216	Enabled	3	-62.2	17.1	63.8	3	-56.9	22.3	63.5
98	9/13/05	2.5	16777216	Enabled	3	-62.2	17.1	69.8	3	-56.9	22.3	69.8
99	9/13/05	2.5	16777216	Enabled	3	-62.2	17.1	69.8	3	-56.9	22.3	70.2
100	9/13/05	2.5	81920000	Enabled	3	-62.2	17.1	68.5	3	-56.9	22.3	68.5
101	9/13/05	2.5	81920000	Enabled	3	-62.2	17.1	70.7	3	-56.9	22.3	70.6
102	9/13/05	2.5	81920000	Enabled	3	-62.2	17.1	70.5	3	-56.9	22.3	70.4
103	9/13/05	2.5	81920000	Enabled	3	-62.2	17.1	70.0	3	-56.9	22.3	70.1
104	9/13/05	2.5	81920000	Enabled	3	-62.2	17.1	70.4	3	-56.9	22.3	70.4
105	9/13/05	2.0	16777216	Enabled	3	-58.3	19.3	72.5	3	-53.3	22.9	72.5
106	9/13/05	2.0	16777216	Enabled	3	-58.3	19.3	69.1	3	-53.3	22.9	69.1
107	9/13/05	2.0	16777216	Enabled	3	-58.3	19.3	69.4	3	-53.3	22.9	69.1
108	9/13/05	2.0	16777216	Enabled	3	-58.3	19.3	70.2	3	-53.3	22.9	70.2
109	9/13/05	2.0	16777216	Enabled	3	-58.3	19.3	70.5	3	-53.3	22.9	70.5
110	9/13/05	2.0	81920000	Enabled	3	-58.3	19.3	69.3	3	-53.3	22.9	69.3
111	9/13/05	2.0	81920000	Enabled	3	-58.3	19.3	72.6	3	-53.3	22.9	72.6
112	9/13/05	2.0	81920000	Enabled	3	-58.3	19.3	72.6	3	-53.3	22.9	72.6
113	9/13/05	2.0	81920000	Enabled	3	-58.3	19.3	72.5	3	-53.3	22.9	72.6
114	9/13/05	2.0	81920000	Enabled	3	-58.3	19.3	72.2	3	-53.3	22.9	72.2
115	9/13/05	1.5	16777216	Enabled	3	-56.3	19.8	71.7	3	-49.4	23.7	72.1
116	9/13/05	1.5	16777216	Enabled	3	-56.3	19.8	71.3	3	-49.4	23.7	71.7
117	9/13/05	1.5	16777216	Enabled	3	-56.3	19.8	76.2	3	-49.4	23.7	76.1
118	9/13/05	1.5	16777216	Enabled	3	-56.3	19.8	64.1	3	-49.4	23.7	64.4
119	9/13/05	1.5	16777216	Enabled	3	-56.3	19.8	75.7	3	-49.4	23.7	76.6
120	9/13/05	1.5	81920000	Enabled	3	-56.3	19.8	71.4	3	-49.4	23.7	71.4
121	9/13/05	1.5	81920000	Enabled	3	-56.3	19.8	72.0	3	-49.4	23.7	72.0
122	9/13/05	1.5	81920000	Enabled	3	-56.3	19.8	75.9	3	-49.4	23.7	75.8
123	9/13/05	1.5	81920000	Enabled	3	-56.3	19.8	67.7	3	-49.4	23.7	67.7
124	9/13/05	1.5	81920000	Enabled	3	-56.3	19.8	73.2	3	-49.4	23.7	73.2
125	9/14/05	1.0	16777216	Enabled	3	-49.9	21.1	64.4	3	-43.4	21.3	74.0
126	9/14/05	1.0	16777216	Enabled	3	-49.9	21.1	77.4	3	-43.4	21.3	77.5
127	9/14/05	1.0	16777216	Enabled	3	-49.9	21.1	70.9	3	-43.4	21.3	70.9
128	9/14/05	1.0	16777216	Enabled	3	-49.9	21.1	74.4	3	-43.4	21.3	74.4
129	9/14/05	1.0	16777216	Enabled	3	-49.9	21.1	75.7	3	-43.4	21.3	75.7
130	9/14/05	1.0	81920000	Enabled	3	-49.9	21.1	74.1	3	-43.4	21.3	74.1
131	9/14/05	1.0	81920000	Enabled	3	-49.9	21.1	72.6	3	-43.4	21.3	72.6
132	9/14/05	1.0	81920000	Enabled	3	-49.9	21.1	71.0	3	-43.4	21.3	71.0
133	9/14/05	1.0	81920000	Enabled	3	-49.9	21.1	77.2	3	-43.4	21.3	77.2
134	9/14/05	1.0	81920000	Enabled	3	-49.9	21.1	73.8	3	-43.4	21.3	73.8